BIODIVERSITY: AN INTRODUCTION

Edited by
Pauline Grace T. Cortes

ArclerPress

Biodiversity: An Introduction

Edited by Pauline Grace T. Cortes

Edition 2018

ArclerPress

Published by
Arcler Press LLC
708 3rd Avenue, 6th Floor
New York
NY 10017
United States of America
www.arclerpress.com

ISBN: 978-1-68094-616-1 (Hardback)
Printed in United States of America
Library of Congress Control Number: 2017933216

About the Editor

Pauline Grace T. Cortes

Pauline graduated in 2011 with a degree in Environmental Planning and Management. Her areas of interest are environmental sustainability, biodiversity conservation, and renewable energy. She is currently working for a global conglomerate to promote a sustainable supply chain based on the principles of United Nations Global Compact

Table of Contents

3 AGRICULTURAL BIODIVERSITY

4 ECOLOGICAL EFFECTS OF BIODIVERSITY

5 HABITAT DESTRUCTION

6 CONSERVATION: ETHICAL USE AND PROTECTION OF VALUABLE RESOURCES

7 CONSERVATION BIOLOGY

8 EFFECTS OF GLOBAL WARMING

9 EFFECTS OF CLIMATE CHANGE ON PLANT BIODIVERSITY

List of Abbreviations

ABAs	Access and Benefit Sharing Agreements
AAAS	American Association for the Advancement of Science
ACP	Asian citrus psyllid
ACSF	Atkinson Center for a Sustainable Future
AMOC	Atlantic Meridional Overturning Circulation
CITES	Convention on International Trade in Endangered Species
DAISY	Digital Automated Identification SYstem
DAD-IS	Domestic Animal Diversity Information System
ESUs	Evolutionary Significant Units
CCSP	Climate Change Science Program
CCD	colony collapse disorder
ITQs	individual transferable quotas
IPCC	Intergovernmental Panel on Climate Change
IUCN	International Union for Conservation of Nature
NRC	National Research Council
SNPs	single-nucleotide polymorphisms
MSY	million species years
THC	thermohaline circulation
SARA	Species at Risk Act

Preface

The number and variety of plants, animals, and other organisms that exist is known as biodiversity. It is an essential component of nature and it ensures the survival of human species by providing food, fuel, shelter, medicines, and other resources to mankind. The richness of biodiversity depends on the climatic conditions and area of the region. All species of plants taken together are called flora and about 70,000 species of plants are known to date. All species of animals taken together are called fauna which includes birds, mammals, fish, reptiles, insects, crustaceans, molluscs, etc. Biodiversity is important everywhere; species and habitats in your area as well as those in distant lands all play a role in maintaining healthy ecosystems. We need biodiversity to satisfy basic needs like food, drinking water, fuel, shelter, and medicine. Much of the world's population still uses plants and animals as a primary source of medicine, and in the United States alone, about 57% of the 150 most prescribed drugs have their origins in biodiversity. Ecosystems provide services such as pollination, seed dispersal, climate regulation, water purification, nutrient cycling, and control of agricultural pests. Many flowering plants depend on animals for pollination, and 30% of human crops depend on the free services of pollinators. Over the last century, humans have come to dominate the planet. Ecosystems are being rapidly altered, and the planet is undergoing a massive loss of biodiversity. Biodiversity is unevenly distributed. It varies globally and within regions. The various factors that influence biodiversity include temperature, altitude, precipitation, soils and their relation with other species. For instance, ocean biodiversity is 25 times lesser than terrestrial diversity. Biodiversity also increases its form as it moves from the poles towards the tropics. While the Earth has always experienced changes and extinctions, the current changes are occurring at an unprecedented rate. Still more sobering, most threats to biodiversity are caused by human activity. Thus we can see that biodiversity which is crucial for the well-being of life on earth, is coming under the threat of many factors related to human activities. There is an urgent need to take action to protect the magnificent biodiversity of our planet. We must create economic policies in order to maintain the Earth's biodiversity and take appropriate measures to protect habitats and species.

This book delivers a global perspective on environmental issues while demonstrating the concept which encompasses the many forms of life on earth and their interdependence on one another for survival.

Content Coverage

Chapter One is intended to provide an overview of biodiversity that has become increasingly prominent the study of the spatial distribution of organisms, species, and ecosystems.

Chapter Two is global biodiversity, the measure of biodiversity on planet Earth and total variability of life forms. More than 99 % of all species that ever lived on Earth are estimated to be extinct.

Chapter Three discusses about the agricultural biodiversity, a sub-set of general biodiversity. It includes all forms of life directly relevant to agriculture: rare seed varieties and animal breeds.

Chapters Four and *Five* are ecological effects of biodiversity and habitat destruction, respectively. Habitat destruction is the process in which natural habitat is rendered unable to support the species present in the area. In this process, the organisms that previously used the site are displaced or destroyed, reducing biodiversity.

Chapter Six covers the conservation which is the ethical use and protection of valuable resources, such as trees, minerals, wildlife, water, and others. It focuses on maintaining the natural world in order to protect the sources of resources.

Chapters Seven and *Eight* is a comprehensive approach on conservation biology techniques and global warming effects respectively. Future effects of climate change will vary depending on climate change policies and social development.

Chapter Nine focuses on the effects of climate change on plant biodiversity. Environmental conditions play a key role in defining the function and distribution of plants, in combination with other factors.

CHAPTER 1

Biodiversity

Chapter Contents

Biodiversity refers to the variation of life forms. It can be used to describe the variation of life in a single ecosystem, a geographical region, or an entire planet. Many biologists believe that biodiversity is an important part of sustainability, and that the more biodiverse a region is, the healthier it is. Biodiversity is greater around the equator and less marked at the poles, due to its harsher and demanding environment.

INTRODUCTION

One of the most widely used definitions of biodiversity refers to it in terms of variability within species, between species, and between ecosystems. It is a measure of the variety of organisms present in different ecosystems. This can refer to genetic variation, ecosystem variation, or species variation (number of species) within an area, biome, or planet. Biodiversity is not distributed evenly on Earth. Terrestrial diversity is greater near the equator due to the result of the warm climate and high primary productivity. Marine diversity is highest along the coasts of the Western Pacific, where sea surface temperature is highest and in the mid-latitudinal band in all oceans. There are latitudinal gradients in species diversity. Biodiversity generally tends to cluster in hotspots,and has been increasing through time, but will likely slow down in the future.

The number and variety of plants, animals, and other organisms that exist is known as biodiversity. It is an essential component of nature and it ensures the survival of human species by providing food, fuel, shelter, medicines, and other resources to mankind. The richness of biodiversity depends on the climatic conditions and area of the region. All species of plants taken together are called flora and about 70,000 species of plants are known to date. All species of animals taken together are called fauna which includes birds, mammals, fish, reptiles, insects, crustaceans, molluscs, etc.

Rapid environmental changes cause mass extinctions. More than 99 percent of all species, amounting to over five billion that lived on Earth are estimated to be extinct. Estimates on the number of Earth's current species ranges from 10 million to 14 million, of which about 1.2 million have been documented and over 86 percent have not yet been described. In May 2016, scientists reported that 1 trillion species are estimated to be present on Earth with only one-thousandth of one percent described. The total amount of related DNA base pairs on Earth is estimated at 5.0×10^{37} and weighs 50 billion tonnes. In comparison, the total mass of the biosphere has been estimated to be as much as 4 TtC (trillion tons of carbon). In July 2016, scientists reported identifying a set of 355 genes from the Last Universal Common Ancestor (LUCA) of all organisms living on Earth.

The age of the Earth is about 4.54 billion years old. The earliest undisputed evidence of life on Earth dates at least from 3.5 billion years ago, during the Eoarchean Era after a geological crust started to solidify following the earlier molten Hadean Eon. There are microbial matfossils found in 3.48 billion year old sandstone discovered in Western Australia. Other early physical evidence of a biogenic substance is graphite in 3.7 billion-year-old meta-sedimentary rocks discovered in Western Greenland. In 2015, "remains of biotic life" were found in 4.1 billion-year-old rocks in Western Australia. According to one of the researchers, if this is directly quoted, If life arose relatively quickly on Earth, then it could be common in the universe.

Since life began on Earth, five major mass extinctions and several minor events have led to large and sudden drops in biodiversity. The Phanerozoic eon (the last 540 million years) marked a rapid growth in biodiversity via the Cambrian explosion—a period during which the majority of multicellular phyla first appeared. The next 400 million years included repeated, massive biodiversity losses classified as mass extinction events. In the Carboniferous, rainforest collapse led to a great loss of plant and animal life. The Permian–Triassic extinction 251 million years ago, was the worst; vertebrate recovery took 30 million years. The most recent, the Cretaceous–Paleogene extinction occurred 65 million years ago and has often attracted more attention than others because it resulted to the extinction of the dinosaurs.

The period since the emergence of humans has displayed an ongoing biodiversity reduction and an accompanying loss of genetic diversity. Named the Holocene extinction, the reduction is caused primarily by human impacts, particularly habitat destruction. Conversely, biodiversity impacts human health in a number of ways, both positively and negatively.

The United Nations designated 2011–2020 as the United Nations Decade on Biodiversity.

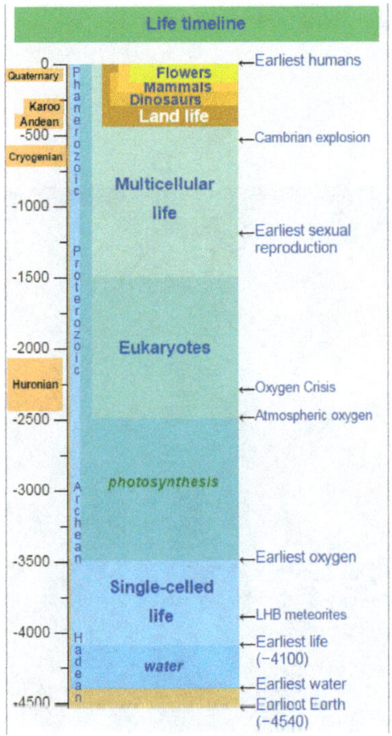

Axis scale: millions of years.

Etymology

The term biological diversity was first used in 1968 by wildlife scientist and conservationist Raymond F. Dasmann in his lay book that advocates conservation, A Different Kind of Country. The term was widely adopted only after more than a decade, when in the 1980s it came into common usage in science and environmental policy. Thomas Lovejoy, in the foreword to the book *Conservation Biology*, introduced the term to the scientific community. Until then, the term "natural diversity" was common, which was introduced by The Science Division of The Nature Conservancy in an important 1975 study, "The Preservation of Natural Diversity." By the early 1980s TNC's Science program and its head, Robert E. Jenkins, Lovejoy, and other leading conservation scientists at the time in America advocated the use of the term "biological diversity".

The term's contracted form *biodiversity* may have been coined by W.G. Rosen in 1985 while planning the 1986 *National Forum on Biological Diversity* organized by the National Research Council (NRC). It first appeared in a publication in 1988 when sociobiologist E. O. Wilson used it as the title of the proceedings of that forum.

Since this period, the term has achieved widespread use among biologists, environmentalists, political leaders, and concerned citizens.

A similar term in the United States is *"natural heritage."* It pre-dates the others and is more accepted by the wider audience interested in conservation. Broader than biodiversity, it includes geology and landforms.

Definitions

A sampling of fungi collected during summer 2008 in Northern Saskatchewan mixed woods, near LaRonge is an example regarding the species diversity of fungi. In this photo, there are also leaf lichens and mosses.

"Biodiversity" is most commonly used to replace the more clearly defined and long established terms, species diversity and species richness. Biologists most often define biodiversity as the "totality of genes, species, and ecosystems of a region". An advantage of this definition is that it seems to describe most circumstances and presents a unified view of the traditional types of biological variety previously identified:

- taxonomic diversity usually measured at the species diversity level
- ecological diversity often viewed from the perspective of ecosystem diversity
- morphological diversity which stems from genetic diversity
- functional diversity which is a measure of the number of functionally disparate species within a population (e.g. different feeding mechanism, different motility, predator vs prey, etc.)

In 2003, Professor Anthony Campbell of Cardiff University, UK and the Darwin Centre, Pembrokeshire, defined a fourth level: Molecular Diversity.

This multilevel construct is consistent with Datman and Lovejoy. An explicit definition consistent with this interpretation was first given in a paper by Bruce A. Wilcox commissioned by the International Union for the Conservation of Nature and Natural Resources (IUCN) for the 1982 World National Parks Conference. Wilcox's definition was "Biological diversity is the variety of life forms...at all levels of biological systems (i.e., molecular, organismic, population, species and ecosystem)...". The 1992 United Nations Earth Summit defined "biological diversity" as "the variability among living organisms from all sources, including, 'inter alia', terrestrial, marine and other aquatic ecosystems, and the ecological complexes of which they are part of: this includes diversity within species, between, species, and of ecosystems". This definition is used in the United Nations Convention on Biological Diversity.

One textbook definition is "variation of life at all levels of biological organization".

Genetically, biodiversity can be defined as the diversity of alleles, genes, and organisms. They study processes such as mutation and gene transfer that drive evolution.

Measuring diversity at one level in a group of organisms may not precisely correspond to diversity at other levels. However, tetrapod (terrestrial vertebrates), taxonomic and ecological diversity shows a very close correlation.

Distribution

A conifer forest in the Swiss Alps (National Park)

Biodiversity is not evenly distributed, rather it varies greatly across the globe as well as within regions. Among other factors, the diversity of all living things (biota) depends on temperature, precipitation, altitude, soils, geography, and the presence of other species. The study of the spatial distribution of organisms, species and ecosystems, is the science of biogeography. Diversity consistently measures higher in the tropics and in other localized regions such as the Cape Floristic Region and lower in polar regions Rain forests that have had wet climates for a long time, such as Yasuní National Park in Ecuador, have particularly high biodiversity. Terrestrial biodiversity is thought to be up to 25 times greater than ocean biodiversity. A recently discovered method put the total number of species on Earth at 8.7 million, of which 2.1 million were estimated to live in the ocean. However, this estimate seems to under-represent the diversity of microorganisms.

Latitudinal gradients

There is an increase in biodiversity from the poles to the tropics. Thus localities at lower latitudes have more species than localities at higher latitudes. This is often referred to as the latitudinal gradient in species diversity. Several ecological mechanisms may contribute to the gradient, but the ultimate factor behind many of them is the greater mean temperature at the equator compared to that of the poles.

Even though terrestrial biodiversity declines from the equator to the poles, some studies claim that this characteristic is unverified in aquatic ecosystems, especially in marine ecosystems. The latitudinal distribution of parasites does not appear to follow this rule.

Hotspots

A biodiversity hotspot is a region with a high level of endemic species that has experienced great habitat loss. The term hotspot was introduced in 1988 by Norman Myers. While hotspots are spread all over the world, the majority are forest areas and most are located in the tropics. Brazil's Atlantic Forest is considered one such hotspot, containing roughly 20,000 plant species, 1,350 vertebrates and millions of insects, about half of which occur nowhere else. The island of Madagascar and India are also particularly notable. Colombia is characterized by high biodiversity, with the highest rate of species by area unit worldwide and it has the largest number of endemics (species that are not found naturally anywhere else) of any country. About 10% of the species of the Earth can be found in Colombia, including over 1,900 species of birds, more than in Europe and North America combined. Colombia has 10% of the world's mammals species, 14% of the amphibian species and 18% of the bird species of the world. Madagascar's dry deciduous forests and lowland rainforests possess a high ratio of endemism. Since the island separated from mainland Africa 66 million years ago, many species and ecosystems have evolved independently. Indonesia's 17,000 islands cover 735,355 square miles (1,904,560 km^2) and contain 10% of the world's flowering plants, 12% of mammals and 17% of reptiles, amphibians and birds—along with nearly 240 million people. Many regions of high biodiversity and/or endemism arise from specialized habitats which require unusual adaptations, for example, alpine environments in high mountains, or Northern European peat bogs.

Accurately measuring differences in biodiversity can be difficult. Selection bias amongst researchers may contribute to biased empirical research for modern estimates of biodiversity. In 1768, Rev. Gilbert White succinctly observed of his Selborne, Hampshire *"all nature is so full, that that district produces the most variety which is the most examined."*

Evolution and History

Biodiversity is the result of 3.5 billion years of evolution. The origin of life has not been definitely established by science, however some evidence suggests that life may already have been well-established only a few hundred million years after the formation of the Earth. Until approximately 600 million years ago, all life consisted of archaea, bacteria, protozoans and similar single-celled organisms.

The history of biodiversity during the Phanerozoic (the last 540 million years), starts with rapid growth during the Cambrian explosion—a period during which nearly every phylum of multicellular organisms first appeared. Over the next 400 million years or so, invertebrate diversity showed little overall trend and vertebrate diversity shows an overall exponential trend. This dramatic rise in diversity was marked by periodic, massive losses of diversity classified as mass extinction events. A significant loss occurred when rainforests collapsed in the carboniferous. The worst was the Permian-Triassic extinction event, 251 million years ago. Vertebrates took 30 million years to recover from this event.

The fossil record suggests that the last few million years featured the greatest biodiversity in history. However, not all scientists support this view, since there is uncertainty as to how strongly the fossil record is biased by the greater availability and preservation of recent geologic sections. Some scientists believe that corrected for sampling artifacts, modern biodiversity may not be much different from biodiversity 300 million years ago. While, others consider the fossil record reasonably reflective of the diversification of life. Estimates of the present global macroscopic species diversity vary from 2 million to 100 million, with a best estimate of somewhere near 9 million where the vast majority are arthropods. Diversity appears to increase continually in the absence of natural selection.

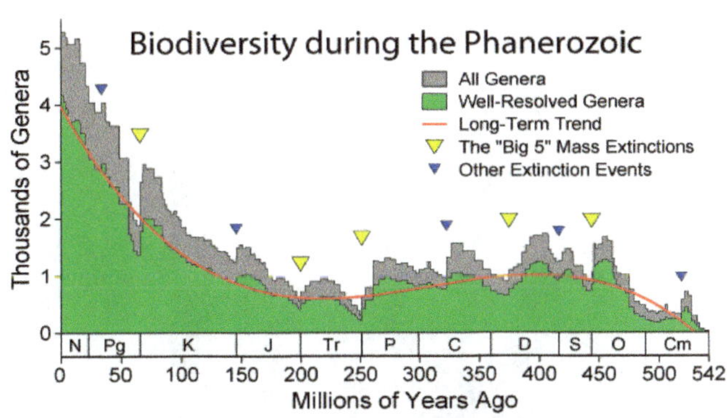

Apparent marine fossil diversity during the Phanerozoic

Evolutionary Diversification

The existence of a "global carrying capacity", limiting the amount of life that can live at once, is debated, as is the question of whether such a limit would also cap the number of species. While records of life in the sea shows a logistic pattern of growth, life on land (insects, plants and tetrapods) shows an exponential rise in diversity. As one author states, "Tetrapods have not yet invaded 64 percent of potentially habitable modes and it could be that without human influence the ecological and taxonomic diversity of tetrapods would continue to increase in an exponential fashion until most or all of the available ecospace is filled."

On the other hand, changes through the Phanerozoic correlate much better with the hyperbolic model (widely used in population biology, demography, and macrosociology, as well as fossil biodiversity) than with exponential and logistic models. The latter models imply that changes in diversity are guided by a first-order positive feedback (more ancestors, more descendants) and/or a negative feedback arising from resource limitation. Hyperbolic model implies a second-order positive feedback. The hyperbolic pattern of the world population growth arises from a second-order positive feedback between the population size and the rate of technological growth. The hyperbolic character of biodiversity growth can be similarly accounted for by a feedback between diversity and community structure complexity. The similarity between the curves of biodiversity and human population probably comes from the fact that both are derived from the interference of the hyperbolic trend with cyclical and stochastic dynamics.

Most biologists agree however that the period since human emergence is part of a new mass extinction, named the Holocene extinction event, caused primarily by the impact humans are having on the environment. It has been argued that the present rate of extinction is sufficient to eliminate most species on the planet Earth within 100 years.

New species are regularly discovered (on average between 5–10,000 new species each year, most of them insects) and many, though discovered, are not yet classified (estimates are that nearly 90% of all arthropods are not yet classified). Most of the terrestrial diversity is found in tropical forests and in general, land has more species than the ocean; some 8.7 million species may exist on Earth, of which some 2.1 million live in the ocean.

Biodiversity and Ecosystem Services

Summer field in Belgium (Hamois). The blue flowers are *Centaurea cyanus* and the red are *Papaver rhoeas*.

The Balance of Evidence

"Ecosystem services are the suite of benefits that ecosystems provide to humanity."

These services come in three flavors:

- Provisioning services which involve the production of renewable resources (e.g.: food, wood, fresh water)
- Regulating services which are those that lessen environmental change (e.g.: climate regulation, pest/disease control)
- Cultural services represent human value and enjoyment (e.g.: landscape aesthetics, cultural heritage, outdoor recreation and spiritual significance)

There have been many claims about biodiversity's effect on these ecosystem services, especially provisioning and regulating services. After an exhaustive survey through peer-reviewed literature to evaluate 36 different claims about biodiversity's effect on ecosystem services, 14 of those claims have been validated, 6 demonstrate mixed support or are unsupported, 3 are incorrect and 13 lack enough evidence to draw definitive conclusions.

Services Enhanced by Biodiversity

Provisioning Services

- Greater species diversity of plants increases fodder yield (synthesis of 271 experimental studies).
- Greater genetic diversity of plants (i.e.: diversity within a single species) increases overall crop yield (synthesis of 575 experimental studies). Although another review of 100 experimental studies reports mixed evidence.
- Greater species diversity of trees increases overall wood production (Synthesis of 53 experimental studies). However, there is not enough data to draw a conclusion about the effect of tree trait diversity on wood production.

Regulating Services

- Greater species diversity of fish increases the stability of fisheries yield (Synthesis of 8 observational studies).
- Greater species diversity of natural pest enemies decreases herbivorous pest populations (Data from two separate reviews; Synthesis of 266 experimental and observational studies; Synthesis of 18 observational studies. Although another review of 38 experimental studies found mixed support for this claim, suggesting that in cases where mutual intraguild predation occurs, a single predatory species is often more effective
- Greater species diversity of plants decreases disease prevalence on plants (Synthesis of 107 experimental studies).
- Greater species diversity of plants increases resistance to plant invasion (Data from two separate reviews; Synthesis of 105 experimental studies; Synthesis of 15 experimental studies).

- Greater species diversity of plants increases carbon sequestration, but note that this finding only relates to actual uptake of carbon dioxide and not long term storage, see below; Synthesis of 479 experimental studies)
- Greater species diversity of plants increases soil nutrient remineralization (Synthesis of 103 experimental studies)
- Greater species diversity of plants increases soil organic matter (Synthesis of 85 experimental studies)

Services with Mixed Evidence

Provisioning Services

- None to date

Regulating Services

- Greater species diversity of plants may or may not decrease herbivorous pest populations. Data from two separate reviews suggest that greater diversity decreases pest populations (Synthesis of 40 observational studies; Synthesis of 100 experimental studies). One review found mixed evidence (Synthesis of 287 experimental studies), while another found contrary evidence (Synthesis of 100 experimental studies).
- Greater species diversity of animals may or may not decrease disease prevalence on those animals (Synthesis of 45 experimental and observational studies), although a 2013 study offers more support showing that biodiversity may in fact enhance disease resistance within animal communities, at least in amphibian frog ponds. Many more studies must be published in support of diversity to sway the balance of evidence will be such that we can draw a general rule on this service.
- Greater species and trait diversity of plants may or may not increase long term carbon storage (Synthesis of 33 observational studies).
- Greater pollinator diversity may or may not increase pollination (Synthesis of 7 observational studies), but a publication from March 2013 suggests that increased native pollinator diversity enhances pollen deposition (although not necessarily fruit set as the authors would have you believe, for details explore their lengthy supplementary material).

Services for which Biodiversity is a Hindrance

Provisioning Services

- Greater species diversity of plants reduces primary production (Synthesis of 7 experimental studies)

Regulating Services

- Greater genetic and species diversity of a number of organisms reduces freshwater purification (Synthesis of 8 experimental studies, although an attempt by the authors to investigate the effect of detritivore diversity on freshwater purification was unsuccessful due to a lack of available evidence, (only 1 observational study was found

Provisioning Services

- Effect of species diversity of plants on biofuel yield (In a survey of the literature, the investigators only found 3 studies).
- Effect of species diversity of fish on fishery yield (In a survey of the literature, the investigators only found 4 experimental studies and 1 observational study).

Regulating Services

- Effect of species diversity on the stability of biofuel yield (In a survey of the literature, the investigators did not find any studies).
- Effect of species diversity of plants on the stability of fodder yield (In a survey of the literature, the investigators only found 2 studies).
- Effect of species diversity of plants on the stability of crop yield (In a survey of the literature, the investigators only found 1 study).
- Effect of genetic diversity of plants on the stability of crop yield (In a survey of the literature, the investigators only found 2 studies).
- Effect of diversity on the stability of wood production (In a survey of the literature, the investigators could not find any studies).
- Effect of species diversity of multiple taxa on erosion control (In a survey of the literature, the investigators could not find any studies – they did however find studies on the effect of species diversity and root biomass).
- Effect of diversity on flood regulation (In a survey of the literature, the investigators could not find any studies).
- Effect of species and trait diversity of plants on soil moisture (In a survey of the literature, the investigators only found 2 studies).

Other sources have reported somewhat conflicting results and in 1997, Robert Costanza and colleagues reported the estimated global value of ecosystem services (not captured in traditional markets) at an average of $33 trillion annually.

Since the stone age, species loss has accelerated above the average basal rate, driven by human activity. Estimates of species losses are at a rate 100-10,000 times as fast as is typical in the fossil record. Biodiversity also affords many non-material benefits including spiritual and aesthetic values, knowledge systems and education.

Biodiversity and Agriculture

Amazon Rainforest in South America

Agricultural diversity can be divided into two categories: Intraspecific diversity refers to the genetic variety within a single species, like the potato (Solanum tiberosum). It is composed of many different forms and types. In the US, we might compare russet potatoes with new potatoes or purple potatoes, all different, but all part of the same species, S. tuberosum.

Interspecific diversity refers to the number and types of different species. Thinking about this diversity, we might note that many small vegetable farmers grow many different crops like potatoes and also carrots, peppers, lettuce, etc.

Agricultural diversity can also be divided by whether it is 'planned' diversity or 'associated' diversity. This is a functional classification that we impose and not an intrinsic feature of life or diversity. Planned diversity includes the crops which a farmer has encouraged, planted, or raised (e.g.: crops, covers, symbionts and livestock, among others), which can be contrasted with the associated diversity that arrives among the crops, uninvited (e.g.: herbivores, weed species and pathogens, among others).

The control of associated biodiversity is one of the great agricultural challenges that farmers face. On monoculture farms, the approach is generally to eradicate associated diversity using a suite of biologically destructive pesticides, mechanized tools and transgenic engineering techniques, then to rotate crops. Although some polyculture farmers use the same techniques, they also employ integrated pest management strategies as well as strategies that are more labor-intensive, but less dependent on capital, biotechnology and energy.

Interspecific crop diversity is, in part, responsible for offering variety in what we eat. Intraspecific diversity, the variety of alleles within a single species, also offers us choice in our diets. If a crop fails in a monoculture, we rely on agricultural diversity to replant the land with something new. If a wheat crop is destroyed by a pest we may plant a hardier variety of wheat the next year, relying on intraspecific diversity. We may forgo wheat production in that area and plant a different species altogether, relying on interspecific diversity. Even an agricultural society which primarily grows monocultures, relies on biodiversity at some point.

- The Irish potato blight of 1846 was a major factor in the deaths of one million people and the emigration of about two million. It was the result of planting only two potato varieties, both vulnerable to the blight, *Phytophthora infestans*, which arrived in 1845.
- When rice grassy stunt virus struck rice fields from Indonesia to India in the 1970s, 6,273 varieties were tested for resistance. Only one was resistant, an Indian variety and known to science only since 1966. This variety formed a hybrid with other varieties and is now widely grown.
- Coffee rust attacked coffee plantations in Sri Lanka, Brazil and Central America in 1970. A resistant variety was found in Ethiopia. The diseases are themselves a form of biodiversity.

Monoculture was a contributing factor to several agricultural disasters, including the European wine industry collapse in the late 19th century and the US southern corn leaf blight epidemic of 1970.

Although about 80 percent of food supply comes from just 20 kinds of plants, humans use at least 40,000 species. Many people depend on these species for food, shelter, and clothing. Earth's surviving biodiversity provides resources for increasing the range of food and other products suitable for human use, although the present extinction rate shrinks that potential.

Biodiversity and Human Health

The diverse forest canopy on Barro Colorado Island, Panama, yielded this display of different fruits.

Biodiversity's relevance to human health is becoming an international political issue, as scientific evidence builds on the global health implications of biodiversity loss. This issue is closely linked with the issue of climate change, as many of the anticipated health risks of climate change are associated with changes in biodiversity (e.g. changes in populations and distribution of disease vectors, scarcity of fresh water, impacts on agricultural biodiversity and food resources, etc.) This is because the species most likely to disappear are those that buffer against infectious disease transmission, while surviving species tend to be the ones that increase disease transmission, such as that of West Nile Virus, Lyme disease, and Hantavirus, according to a study done co-authored by Felicia Keesing, an ecologist at Bard College and Drew Harvell, associate director for Environment of the Atkinson Center for a Sustainable Future (ACSF) at Cornell University.

The growing demand and lack of drinkable water on the planet presents an additional challenge to the future of human health. Partly, the problem lies in the success of water suppliers to increase supplies and failure of groups promoting preservation of water resources. While the distribution of clean water increases, in some parts of the world it remains unequal. According to a *2008 World Population Data Sheet*, only 62% of least developed countries are able to access clean water.

Some of the health issues influenced by biodiversity include dietary health and nutrition security, infectious disease, medical science and medicinal resources, social and psychological health. Biodiversity is also known to have an important role in reducing disaster risk and in post-disaster relief and recovery efforts.

Biodiversity provides critical support for drug discovery and the availability of medicinal resources. A significant proportion of drugs are derived, directly or indirectly, from biological sources: at least 50% of the pharmaceutical compounds on the US market are derived from plants, animals and micro-organisms, while about 80% of the world population depends on medicines from nature (used in either modern or traditional medical practice) for primary healthcare. Only a tiny fraction of wild species has been investigated for medical potential. Biodiversity has been critical to advances throughout the field of bionics. Evidence from market analysis and biodiversity science indicates that the decline in output from the pharmaceutical sector since the mid-1980s can be attributed to a move away from natural product exploration ("bioprospecting") in favor of genomics and synthetic chemistry, indeed, claims about the value of undiscovered pharmaceuticals may not provide enough incentive for companies in free markets to search for them because of the high cost of development; meanwhile, natural products have a long history of supporting significant economic and health innovation. Marine ecosystems are particularly important, although inappropriate bioprospecting can increase biodiversity loss, as well as violating the laws of the communities and states from which the resources are taken.

Biodiversity, Business and Industry

Agriculture production, pictured is a tractor and a chaser bin

Many industrial materials derive directly from biological sources. These include building materials, fibers, dyes, rubber and oil. Biodiversity is also important to the security of resources such as water, timber, paper, fiber and food. As a result, biodiversity loss is a significant risk factor in business development and a threat to long term economic sustainability.

Biodiversity, Leisure, Cultural, and Aesthetic Value

Biodiversity enriches leisure activities such as hiking, birdwatching, or natural history study. Biodiversity inspires musicians, painters, sculptors, writers, and other artists. Many cultures view themselves as an integral part of the natural world which requires them to respect other living organisms.

Popular activities such as gardening, fishkeeping, and specimen collecting strongly depend on biodiversity. The number of species involved in such pursuits is in the tens of thousands, though the majority do not enter commerce.

The relationships between the original natural areas of these often exotic animals and plants and commercial collectors, suppliers, breeders, propagators and those who promote their understanding and enjoyment are complex and poorly understood. The general public responds well to exposure to rare and unusual organisms, reflecting their inherent value.

Philosophically, it could be argued that biodiversity has intrinsic aesthetic and spiritual value to mankind *in and of itself*. This idea can be used as a counterweight to the notion that tropical forests and other ecological realms are only worthy of conservation because of the services they provide.

Biodiversity and Ecological Services

Eagle Creek, Oregon hiking

Biodiversity supports many ecosystem services:

"There is now unequivocal evidence that biodiversity loss reduces the efficiency by which ecological communities capture biologically essential resources, produce biomass, decompose and recycle biologically essential nutrients... There is mounting evidence that biodiversity increases the stability of ecosystem functions through time... Diverse communities are more productive because they contain key species that have a large influence on productivity and differences in functional traits among organisms increase total resource capture... The impacts of diversity loss on ecological processes might be sufficiently large to rival the impacts of many other global drivers of environmental change... Maintaining multiple ecosystem processes at multiple places and times requires higher levels of biodiversity than does a single process at a single place and time."

It plays a part in regulating the chemistry of our atmosphere and water supply. Biodiversity is directly involved in water purification, recycling nutrients and providing fertile soils. Experiments with controlled environments have shown that humans cannot easily build ecosystems to support human needs; for example insect pollination cannot be mimicked, and that activity alone represented between $2.1-14.6 billions in 2003.

Number of Species

Species	Earth			Ocean		
	Catalogued	Predicted	±SE	Catalogued	Predicted	±SE
Eukaryotes						
Animalia	953,434	7,770,000	958,000	171,082	2,150,000	145,000
Chromista	13,033	27,500	30,500	4,859	7,400	9,640
Fungi	43,271	611,000	297,000	1,097	5,320	11,100
Plantae	215,644	298,000	8,200	8,600	16,600	9,130
Protozoa	8,118	36,400	6,690	8,118	36,400	6,690
Total	1,233,500	8,740,000	1,300,000	193,756	2,210,000	182,000
Prokaryotes						
Archaea	502	455	160	1	1	0
Bacteria	10,358	9,680	3,470	652	1,320	436
Total	10,860	10,100	3,630	653	1,320	436
Grand Total	**1,244,360**	**8,750,000**	**1,300,000**	**194,409**	**2,210,000**	**182,000**

Predictions for prokaryotes represent a lower bound because they do not consider undescribed higher taxa. For protozoa, the ocean database was substantially more complete than the database for the entire Earth so we only used the former to estimate the total number of species in this taxon. All predictions were rounded to three significant digits.
doi:10.1371/journal.pbio.1001127.t002

According to Mora and colleagues, the total number of terrestrial species is estimated to be around 8.7 million, while the number of oceanic species is much lower, estimated at 2.2 million. The authors note that these estimates are strongest for eukaryotic organisms and likely represent the lower bound of prokaryote diversity. Other estimates include:

- 220,000 vascular plants, estimated using the species-area relation method
- 0.7-1 million marine species
- 10–30 million insects; (of some 0.9 million we know today)
- 5–10 million bacteria;
- 1.5-3 million fungi, estimates based on data from the tropics, long-term non-tropical sites and molecular studies that have revealed cryptic speciation. Some 0.075 million species of fungi had been documented by 2001).
- 1 million mites
- The number of microbial species is not reliably known, but the Global Ocean Sampling Expedition dramatically increased the estimates of genetic diversity by identifying an enormous number of new genes from near-surface plankton samples at various marine locations, initially over the 2004-2006 period. The findings may eventually cause a significant change in the way science defines species and other taxonomic categories.

Since the rate of extinction has increased, many extant species may become extinct before they are described. Not surprisingly, in the animalia the most studied groups are birds and mammals, whereas fishes and arthropods are the least studied animals groups.

Species Loss Rates

No longer do we have to justify the existence of humid tropical forests on the feeble grounds that they might carry plants with drugs that cure human disease. *Gaia theory* forces us to see that they offer much more than this. Through their capacity to evapotranspirate vast volumes of water vapor, they serve to keep the planet cool by wearing a sunshade of white reflecting cloud. Their replacement by cropland could precipitate a disaster that is global in scale.

During the last century, decreases in biodiversity have been increasingly observed. In 2007, German Federal Environment Minister Sigmar Gabriel cited estimates that up to 30% of all species will be extinct by 2050. Of these, about one eighth of known plant species are threatened with extinction. Estimates reach as high as 140,000 species per year (based on Species-area theory). This figure indicates unsustainable ecological practices, because few species emerge each year. Almost all scientists acknowledge that the rate of species loss is greater now than at any time in human history, with extinctions occurring at rates hundreds of times higher than background extinction rates. As of 2012, some studies suggest that 25% of all mammal species could be extinct in 20 years.

In absolute terms, the planet has lost 52% of its biodiversity since 1970 according to a 2014 study by the World Wildlife Fund. The Living Planet Report 2014 claims that the number of mammals, birds, reptiles, amphibians and fish across the globe is, on average, about half the size it was 40 years ago. Of that number, 39% accounts for the terrestrial wildlife gone, 39% for the marine wildlife gone and 76% for the freshwater wildlife gone. Biodiversity took the biggest hit in Latin America, plummeting 83 percent. High-income countries showed a 10% increase in biodiversity, which was canceled out by a loss in low-income countries. This is despite the fact that high-income countries use five times the ecological resources of low-income countries, which was explained as a result of process whereby wealthy nations are outsourcing resource depletion to poorer nations, which are suffering the greatest ecosystem losses.

Threats

In 2006 many species were formally classified as rare or endangered or threatened. Moreover, scientists have estimated that millions more species are at risk which have not been formally recognized. About 40 percent of the 40,177 species assessed using the IUCN Red List criteria are now listed as threatened with extinction—a total of 16,119.

Jared Diamond describes an "Evil Quartet" of habitat destruction, overkill, introduced species and secondary extinctions. Edward O. Wilson prefers the acronym HIPPO, standing for **H**abitat destruction, **I**nvasive species, **P**ollution, human over-**P**opulation and **O**ver-harvesting. The most authoritative classification in use today is IUCN's Classification of Direct Threats which has been adopted by major international conservation organizations such as the US Nature Conservancy, the World Wildlife Fund, Conservation International and BirdLife International.

Habitat Destruction

Deforestation and increased road-building in the Amazon Rainforest are a significant concern because of increased human encroachment upon wild areas, increased resource extraction and further threats to biodiversity.

Habitat destruction has played a key role in extinctions, especially related to tropical forest destruction.Factors contributing to habitat loss are: overconsumption, overpopulation, land use change, deforestation, pollution (air pollution, water pollution, soil contamination) and global warming or climate change.

Habitat size and numbers of species are systematically related. Physically larger species and those living at lower latitudes or in forests or oceans are more sensitive to reduction in habitat area. Conversion to "trivial" standardized ecosystems (e.g., monoculture following deforestation) effectively destroys habitat for the more diverse species that preceded the conversion. In some countries lack of property rights or lax law/regulatory enforcement necessarily leads to biodiversity loss (degradation costs having to be supported by the community).

A 2007 study conducted by the National Science Foundation found that biodiversity and genetic diversity are codependent—that diversity among species requires diversity within a species and vice versa. "If any one type is removed from the system, the cycle can break down and the community becomes dominated by a single species." At present, the most threatened ecosystems are found in fresh water, according to the Millennium Ecosystem Assessment 2005, which was confirmed by the "Freshwater Animal Diversity Assessment", organised by the biodiversity platform and the French *Institut de recherche pour le développement* (MNHNP).

Co-extinctions are a form of habitat destruction. Co-extinction occurs when the extinction or decline in one accompanies the other, such as in plants and beetles.

Introduced and Invasive Species

Male *Lophura nycthemera* (silver pheasant), a native of East Asia that has been introduced into parts of Europe for ornamental reasons

Barriers such as large rivers, seas, oceans, mountains and deserts encourage diversity by enabling independent evolution on either side of the barrier, via the process of allopatric speciation. The term invasive species is applied to species that breach the natural barriers that would normally keep them constrained. Without barriers, such species occupy new territory, often supplanting native species by occupying their niches, or by using resources that would normally sustain native species.

The number of species invasions has been on the rise at least since the beginning of the 1900s. Species are increasingly being moved by humans (on purpose and accidentally). In some cases the invaders are causing drastic changes and damage to their new habitats (e.g.: zebra mussels and the emerald ash borer in the Great Lakes region and the lion fish along the North American Atlantic coast). Some evidence suggests that invasive species are competitive in their new habitats because they are subject to less pathogen disturbance. Others report confounding evidence that occasionally suggest that species-rich communities harbor many native and exotic species simultaneously while some say that diverse ecosystems are more resilient and resist invasive plants and animals. An important question is, do invasive species cause extinctions? Many studies cite effects of invasive species on natives, but not extinctions. Invasive species seem to increase local (i.e.: alpha diversity) diversity, which decreases turnover of diversity (i.e.: beta diversity). Overall gamma diversity may be lowered because species are going extinct because of other causes, but even some of the most insidious invaders (e.g.: Dutch elm disease, emerald ash borer, chestnut blight in North America) have not caused their host species to become extinct. Extirpation, population decline and homogenization of regional biodiversity are much more common. Human activities have frequently been the cause of invasive species circumventing their barriers, by introducing them for food and other purposes. Human activities therefore allow species to migrate to new areas (and thus become invasive) occurred on time scales much shorter than historically have been required for a species to extend its range.

Not all introduced species are invasive, nor all invasive species deliberately introduced. In cases such as the zebra mussel, invasion of US waterways was unintentional. In other cases, such as mongooses in Hawaii, the introduction is deliberate but ineffective (nocturnal rats were not vulnerable to the diurnal mongoose). In other cases, such as oil palms in Indonesia and Malaysia, the introduction produces substantial economic benefits, but the benefits are accompanied by costly unintended consequences.

Finally, an introduced species may unintentionally injure a species that depends on the species it replaces. In Belgium, Prunus spinosa from Eastern Europe leafs much sooner than its West European counterparts, disrupting the feeding habits of the *Thecla betulae* butterfly (which feeds on the leaves). Introducing new species often leaves endemic and other local species unable to compete with the exotic species and unable to survive. The exotic organisms may be predators, parasites, or may simply outcompete indigenous species for nutrients, water and light.

At present, several countries have already imported so many exotic species, particularly agricultural and ornamental plants, that their own indigenous fauna/flora may be outnumbered. For example, the introduction of kudzu from Southeast Asia to Canada and the United States has threatened biodiversity in certain areas.

Genetic Pollution

Endemic species can be threatened with extinction through the process of genetic pollution, i.e. uncontrolled hybridization, introgression and genetic swamping. Genetic pollution leads to homogenization or replacement of local genomes as a result of either a numerical and/ or fitness advantage of an introduced species. Hybridization and introgression are side-effects of introduction and invasion. These phenomena can be especially detrimental to rare species that come into contact with more abundant ones. The abundant species can interbreed with the rare species, swamping its gene pool. This problem is not always apparent from morphological (outward appearance) observations alone. Some degree of gene flow is normal adaptation and not all gene and genotype constellations can be preserved. However, hybridization with or without introgression may, nevertheless, threaten a rare species' existence.

Overexploitation

Overexploitation occurs when a resource is consumed at an unsustainable rate. This occurs on land in the form of overhunting, excessive logging, poor soil conservation in agriculture and the illegal wildlife trade.

About 25% of world fisheries are now overfished to the point where their current biomass is less than the level that maximizes their sustainable yield.

The overkill hypothesis, a pattern of large animal extinctions connected with human migration patterns, can be used to explain why megafaunal extinctions can occur within a relatively short time period.

Hybridization, Genetic Pollution/Erosion and Food Security

The Yecoro wheat (right) cultivar is sensitive to salinity, plants resulting from a hybrid cross with cultivar W4910 (left) show greater tolerance to high salinity.

In agriculture and animal husbandry, the Green Revolution popularized the use of conventional hybridization to increase yield. Often hybridized breeds originated in developed countries and were further hybridized with local varieties in the developing world to create high yield strains resistant to local climate and diseases. Local governments and industry have been pushing hybridization. Formerly huge gene pools of various wild and indigenous breeds have collapsed causing widespread genetic erosion and genetic pollution. This has resulted to loss of genetic diversity and biodiversity as a whole.

(GM organisms) have genetic material altered by genetic engineering procedures such as recombinant DNA technology. GM crops have become a common source for genetic pollution, not only of wild varieties but also of domesticated varieties derived from classical hybridization.

Genetic erosion coupled with genetic pollution may be destroying unique genotypes, thereby creating a hidden crisis which could result a severe threat to our food security. Diverse genetic material could cease to exist which would impact our ability to further hybridize food crops and livestock against more resistant diseases and climatic changes.

Climate Change

Polar bears on the sea ice of the Arctic Ocean, near the North Pole. Climate change has started affecting bear populations.

Global warming is also considered to be a major potential threat to global biodiversity in the future.For example, coral reefs - which are biodiversity hotspots - will be lost within the century if global warming continues at the current trend.

Climate change has seen many claims about potential to affect biodiversity but evidence supporting the statement is tenuous. Increasing atmospheric carbon dioxide certainly affects plant morphology and is acidifying oceans, and temperature affects species ranges, phenology, and weather, but the major impacts that have been predicted are still just *potential* impacts. We have not documented major extinctions yet, even as climate change drastically alters the biology of many species.

In 2004, an international collaborative study on four continents estimated that 10 percent of species would become extinct by 2050 because of global warming. We need to limit climate change or we wind up with a lot of species in trouble, possibly extinct, said Dr. Lee Hannah, a co-author of the paper and chief climate change biologist at the Center for Applied Biodiversity Science at Conservation International.

A recent study predicts that up to 35% of the world terrestrial carnivores and ungulates will be at higher risk of extinction by 2050 because of the joint effects of predicted climate and land-use change under business-as-usual human development scenarios.

Human Overpopulation

From 1950 to 2011, world population increased from 2.5 billion to 7 billion and is forecast to reach a plateau of more than 9 billion during the 21st century. Some recent forecasts placed the possible number of people on the planet at 11 billion or 15 billion by 2100. Sir David King, former chief scientific adviser to the UK government, told a parliamentary inquiry: *"It is self-evident that the massive growth in the human population through the 20th century has had more impact on biodiversity than any other single factor."* At least until the middle of the 21st century, worldwide losses of pristine biodiverse land will probably depend much on the worldwide human birth rate.

According to a 2014 study by the World Wildlife Fund, the global human population already exceeds planet's biocapacity - it would take the equivalent of 1.5 Earths of biocapacity to meet our current demands. The report further points that if everyone on the planet had the footprint of the average resident of Qatar, we would need 4.8 Earths and if we lived the lifestyle of a typical resident of the USA, we would need 3.9 Earths.

The Holocene Extinction

Rates of decline in biodiversity in this sixth mass extinction match or exceed rates of loss in the five previous mass extinction events in the fossil record. Loss of biodiversity results in the loss of natural capital that supplies ecosystem goods and services. From the perspective of the method known as Natural Economy, the economic value of 17 ecosystem services for Earth's biosphere (calculated in 1997) has an estimated value of US$33 trillion ($3.3x10^{13}$) per year.

Conservation

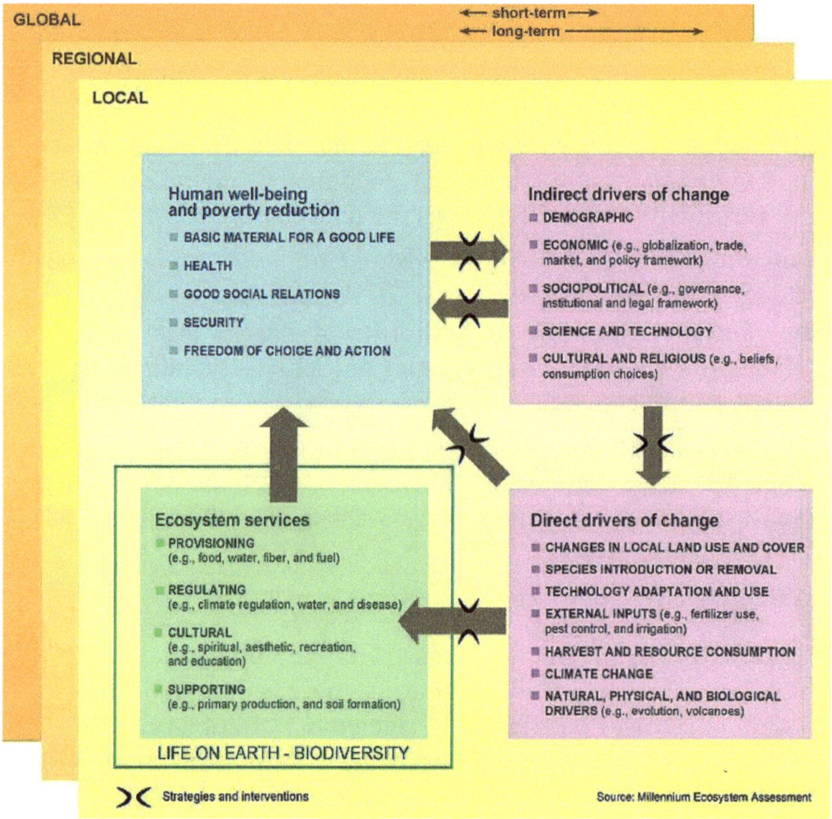

A schematic image illustrating the relationship between biodiversity, ecosystem services, human well-being, and poverty. The illustration shows where conservation action, strategies, and plans can influence the drivers of the current biodiversity crisis at local, regional, to global scales.

The retreat of Aletsch Glacier in the Swiss Alps (situation in 1979, 1991, and 2002), due to global warming.

Conservation biology matured in the mid-20th century as ecologists, naturalists, and other scientists began to research and address issues pertaining to global biodiversity declines.

The conservation ethic advocates management of natural resources for the purpose of sustaining biodiversity in species, ecosystems, the evolutionary process and human culture and society.

Conservation biology is reforming around strategic plans to protect biodiversity. Preserving global biodiversity is a priority in strategic conservation plans that are designed to engage public policy and concerns affecting local, regional and global scales of communities, ecosystems and cultures. Action plans identify ways of sustaining human well-being, employing natural capital, market capital, and ecosystem services.

In the EU Directive 1999/22/EC, zoos are described as having a role in the preservation of the biodiversity of wildlife animals by conducting research or participation in breeding programs.

Protection and Restoration Techniques

Removal of exotic species will allow the species that they have negatively impacted to recover their ecological niches. Exotic species that have become pests can be identified taxonomically (e.g., with Digital Automated Identification SYstem (DAISY), using the barcode of life). Removal is practical only given large groups of individuals due to the economic cost.

As sustainable populations of the remaining native species in an area become assured, missing species that are candidates for reintroduction can be identified using databases such as the *Encyclopedia of Life* and the Global Biodiversity Information Facility.

- Biodiversity banking places a monetary value on biodiversity. One example is the Australian Native Vegetation Management Framework.
- Gene banks are collections of specimens and genetic material. Some banks intend to reintroduce banked species to the ecosystem (e.g., via tree nurseries).
- Reduction of and better targeting of pesticides allows more species to survive in agricultural and urbanized areas.
- Location-specific approaches may be less useful for protecting migratory species. Onc approach is to create wildlife corridors that correspond to the animals' movements. National and other boundaries can complicate corridor creation.

Protected Areas

Protected areas are meant for affording protection of wild animals and their habitat, This also includes forest reserves and biosphere reserves. Protected areas have been set up all over the world with the specific aim of protecting and conserving plants and animals.

National Parks

National park and nature reserve is the area selected by governments or private organizations for special protection against damage or degradation with the objective of biodiversity and landscape conservation. National parks are usually owned and managed by national or state governments. A limit is placed on the number of visitors permitted to enter certain fragile areas. Designated trails or roads are created. The visitors are allowed to enter only for study, cultural, and recreational purposes. Forestry operations, grazing of animals, and hunting of animals are regulated. Exploitation of habitat or wildlife is banned.

Wildlife Sanctuary

Wildlife sanctuary aims only at conservation of species and have the following features:

- The boundaries of the sanctuaries are not limited by state legislation.
- The killing, hunting or capturing of any species is prohibited except by or under the control of the highest authority in the department which is responsible for the management of the sanctuary.
- Private ownership may be allowed.
- Forestry and other usages can also be permitted.

Forest Reserves

The forests play a vital role in harbouring more than 45,000 floral and 81,000 faunal species of which 5,150 floral and 1,837 faunal species are endemic. Plant and animal species confined to a specific geographical area are called endemic species. In reserved forests, rights to activities like hunting and grazing are sometimes given to communities living on the fringes of the forest, who sustain their livelihood partially or wholly from forest resources or products. The unclassed forests covers 6.4 percent of the total forest area and they are marked by the following characteristics:

- They are large inaccessible forests.
- Many of these are unoccupied.
- They are ecologically and economically less important.

Steps to Conserve the Forest Cover

- An extensive reforestation/afforestation program should be followed.
- Alternative environment-friendly sources of fuel energy such as biogas other than wood should be used.
- Loss of biodiversity due to forest fire is a major problem, immediate steps to prevent forest fire need to be taken.
- Overgrazing by cattle can damage a forest seriously. Therefore, certain steps should be taken to prevent overgrazing by cattle.
- Hunting and poaching should be banned.

Zoological Parks

In zoological parks or zoos, live animals are kept for public recreation, education, and conservation purposes. Modern zoos offer veterinary facilities, provide opportunities for threatened species to breed in captivity and usually build environments that simulate the native habitats of the animals in their care. Zoos play a major role in creating awareness among common people about the need to conserve nature.

Botanical Gardens

Botanical garden is a garden in which plants are grown and displayed primarily for scientific and educational purposes. It consists of a collection of living plants, grown outdoors or under glass in greenhouses and conservatories. In addition, it includes a collection of dried plants or herbarium and such facilities as lecture rooms, laboratories, libraries, museums, and experimental or research plantings.

Resource Allocation

Focusing on limited areas of higher potential biodiversity promises greater immediate return on investment than spreading resources evenly or focusing on areas of little diversity but greater interest in biodiversity.

A second strategy focuses on areas that retain most of their original diversity, which typically require little or no restoration. These are typically non-urbanized, non-agricultural areas. Tropical areas often fit both criteria, given their natively high diversity and relative lack of development.

Legal Status

A great deal of work is carried out to preserve the natural characteristics of Hopetoun Falls, Australia while continuing to allow visitor access.

International

- United Nations Convention on Biological Diversity (1992) and Cartagena Protocol on Biosafety;
- Convention on International Trade in Endangered Species (CITES);
- Ramsar Convention (Wetlands);
- Bonn Convention on Migratory Species;
- World Heritage Convention (indirectly by protecting biodiversity habitats)
- Regional Conventions such as the Apia Convention
- Bilateral agreements such as the Japan-Australia Migratory Bird Agreement.

Global agreements such as the Convention on Biological Diversity, give "sovereign national rights over biological resources" (not property). The agreements commit countries to "conserve biodiversity", "develop resources for sustainability" and "share the benefits" resulting from their use. Biodiverse countries that allow bioprospecting or collection of natural products, expect a share of the benefits rather than allowing the individual or institution that discovers/exploits the resource to capture them privately. Bioprospecting can become a type of biopiracy when such principles are not respected.

Sovereignty principles can rely upon what is better known as Access and Benefit Sharing Agreements (ABAs). The Convention on Biodiversity implies informed consent between the source country and the collector, to establish which resource will be used and for what and to settle on a fair agreement on benefit sharing.

National Level Laws

Biodiversity is taken into account in some political and judicial decisions:

- The relationship between law and ecosystems is very ancient and has consequences for biodiversity. It is related to private and public property rights. It can define protection for threatened ecosystems, but also some rights and duties (for example, fishing and hunting rights).
- Law regarding species is more recent. It defines species that must be protected because they may be threatened by extinction. The U.S. Endangered Species Act is an example of an attempt to address the "law and species" issue.
- Laws regarding gene pools are only about a century old. Domestication and plant breeding methods are not new, but advances in genetic engineering have led to tighter laws covering distribution of genetically modified organisms, gene patents and process patents. Governments struggle to decide whether to focus on for example, genes, genomes, or organisms, and species.

Uniform approval for use of biodiversity as a legal standard has not been achieved, however. Bosselman argues that biodiversity should not be used as a legal standard, claiming that the remaining areas of scientific uncertainty cause unacceptable administrative waste and increase litigation without promoting preservation goals.

India passed the Biological Diversity Act in 2002 for the conservation of biological diversity in India. The Act also provides mechanisms for equitable sharing of benefits from the use of traditional biological resources and knowledge.

Analytical Limits

Taxonomic and Size Relationships

Less than 1% of all species that have been described were studied beyond simply noting their existence. The vast majority of Earth's species are microbial. Contemporary biodiversity physics is "firmly fixated on the visible [macroscopic] world". For example, microbial life is metabolically and environmentally more diverse than multicellular life (see e.g., extremophile). "On the tree of life, based on analyses of small-subunit ribosomal RNA, visible life consists of barely noticeable twigs. The inverse relationship of size and population recurs higher on the evolutionary ladder—"to a first approximation, all multicellular species on Earth are insects". Insect extinction rates are high—supporting the Holocene extinction hypothesis.

REFERENCES

1. "What is biodiversity?" (PDF). United Nations Environment Programme, World Conservation Monitoring Centre.
2. Gaston, Kevin J. (11 May 2000). "Global patterns in biodiversity". Nature. 405 (6783): 220–227. doi:10.1038/35012228. PMID 10821282.
3. Field, Richard; Hawkins, Bradford A.; Cornell, Howard V.; Currie, David J.; Diniz-Filho, J. (1 January 2009). Alexandre F.; Guégan, Jean-François; Kaufman, Dawn M.; Kerr, Jeremy T.; Mittelbach, Gary G.; Oberdorff, Thierry; O'Brien, Eileen M.; Turner, John R. G.. "Spatial species-richness gradients across scales: a meta-analysis". Journal of Biogeography. 36 (1): 132–147. doi:10.1111/j.1365-2699.2008.01963.x.
4. Tittensor, Derek P.; Mora, Camilo; Jetz, Walter; Lotze, Heike K.; Ricard, Daniel; Berghe, Edward Vanden; Worm, Boris (28 July 2010). "Global patterns and predictors of marine biodiversity across taxa". Nature. 466 (7310): 1098–1101. Bibcode:2010Natur.466.1098T. doi:10.1038/nature09329. PMID 20668450.
5. Myers, Norman; Mittermeier, Russell A.; Mittermeier, Cristina G.; Da Fonseca, Gustavo A. B.; Kent, Jennifer (24 February 2000). "Biodiversity hotspots for conservation priorities". Nature. 403 (6772): 853–858. Bibcode:2000Natur.403..853M. doi:10.1038/35002501. PMID 10706275.
6. McPeek, Mark A.; Brown, Jonathan M. (1 April 2007). "Clade Age and Not Diversification Rate Explains Species Richness among Animal Taxa". The American Naturalist. 169 (4): E97–E106. doi:10.1086/512135. PMID 17427118.
7. Peters, Shanan. "Sepkoski's Online Genus Database". University of Wisconsin-Madison. Retrieved 10 April 2013.

8. Rabosky, Daniel L. (1 August 2009). "Ecological limits and diversification rate: alternative paradigms to explain the variation in species richness among clades and regions". Ecology Letters. 12 (8): 735–743. doi:10.1111/j.1461-0248.2009.01333.x. PMID 19558515.

9. Cockell, Charles; Koeberl, Christian; Gilmour, Iain (18 May 2006). Biological Processes Associated with Impact Events (1 ed.). Springer Science & Business Media. pp. 197–219. ISBN 978-3-540-25736-3.

10. Algeo, T. J.; Scheckler, S. E. (29 January 1998). "Terrestrial-marine teleconnections in the Devonian: links between the evolution of land plants, weathering processes, and marine anoxic events". Philosophical Transactions of the Royal Society B: Biological Sciences. 353 (1365): 113–130. doi:10.1098/rstb.1998.0195.

11. Bond, David P.G.; Wignall, Paul B. (1 June 2008). "The role of sea-level change and marine anoxia in the Frasnian–Famennian (Late Devonian) mass extinction". Palaeogeography, Palaeoclimatology, Palaeoecology. 263 (3–4): 107–118. doi:10.1016/j.palaeo.2008.02.015.

12. Kunin, W.E.; Gaston, Kevin, eds. (31 December 1996). The Biology of Rarity: Causes and consequences of rare—common differences. ISBN 978-0412633805. Retrieved 26 May 2015.

13. Stearns, Beverly Peterson; Stearns, S. C.; Stearns, Stephen C. (2000). Watching, from the Edge of Extinction. Yale University Press. p. 1921. ISBN 978-0-300-08469-6. Retrieved 2014-12-27.

14. Novacek, Michael J. (8 November 2014). "Prehistory's Brilliant Future". New York Times. Retrieved 2014-12-25.

15. G. Miller; Scott Spoolman (2012). Environmental Science - Biodiversity Is a Crucial Part of the Earth's Natural Capital. Cengage Learning. p. 62. ISBN 1-133-70787-4. Retrieved 2014-12-27.

16. Mora, C.; Tittensor, D.P.; Adl, S.; Simpson, A.G.; Worm, B. (23 August 2011). "How many species are there on Earth and in the ocean?". PLOS Biology. 9 (8): e1001127. doi:10.1371/journal.pbio.1001127. PMC 3160336 . PMID 21886479.

17. Staff (2 May 2016). "Researchers find that Earth may be home to 1 trillion species". National Science Foundation. Retrieved 6 May 2016.

18. Nuwer, Rachel (18 July 2015). "Counting All the DNA on Earth". The New York Times. New York: The New York Times Company. ISSN 0362-4331. Retrieved 2015-07-18.

19. "The Biosphere: Diversity of Life". Aspen Global Change Institute. Basalt, CO. Retrieved 2015-07-19.

20. Wade, Nicholas (25 July 2016). "Meet Luca, the Ancestor of All Living Things". New York Times. Retrieved 25 July 2016.

21. "Age of the Earth". U.S. Geological Survey. 1997. Archived from the original on 23 December 2005. Retrieved 2006-01-10.

22. Dalrymple, G. Brent (2001). "The age of the Earth in the twentieth century: a problem (mostly) solved". Special Publications, Geological Society of

London. 190 (1): 205–221. Bibcode:2001GSLSP.190..205D. doi:10.1144/GSL.SP.2001.190.01.14.

23. Manhesa, Gérard; Allègre, Claude J.; Dupréa, Bernard & Hamelin, Bruno (1980). "Lead isotope study of basic-ultrabasic layered complexes: Speculations about the age of the earth and primitive mantle characteristics". Earth and Planetary Science Letters. 47 (3): 370–382. Bibcode:1980E&PSL..47..370M. doi:10.1016/0012-821X(80)90024-2.
24. Schopf, J. William; Kudryavtsev, Anatoliy B.; Czaja, Andrew D.; Tripathi, Abhishek B. (2007-10-05). "Evidence of Archean life: Stromatolites and microfossils". Precambrian Research. Earliest Evidence of Life on Earth. 158 (3–4): 141–155. doi:10.1016/j.precamres.2007.04.009.
25. Schopf, J. William (2006-06-29). "Fossil evidence of Archaean life". Philosophical Transactions of the Royal Society B: Biological Sciences. 361 (1470): 869–885. doi:10.1098/rstb.2006.1834. ISSN 0962-8436. PMC 1578735 . PMID 16754604.
26. Hamilton Raven, Peter; Brooks Johnson, George (2002). Biology. McGraw-Hill Education. p. 68. ISBN 978-0-07-112261-0. Retrieved 2013-07-07.
27. Borenstein, Seth (13 November 2013). "Oldest fossil found: Meet your microbial mom". AP News.
28. Pearlman, Jonathan (13 November 2013). "'Oldest signs of life on Earth found' - Scientists discover potentially oldest signs of life on Earth – 3.5 billion-year-old microbe traces in rocks in Australia". The Telegraph. Retrieved 2014-12-15.
29. Noffke, Nora; Christian, Daniel; Wacey, David; Hazen, Robert M. (8 November 2013). "Microbially Induced Sedimentary Structures Recording an Ancient Ecosystem in the ca. 3.48 Billion-Year-Old Dresser Formation, Pilbara, Western Australia". Astrobiology (journal). 13 (12): 1103–24. Bibcode:2013AsBio..13.1103N. doi:10.1089/ast.2013.1030. PMC 3870916 . PMID 24205812.
30. Ohtomo, Yoko; Kakegawa, Takeshi; Ishida, Akizumi; Nagase, Toshiro; Rosing, Minik T. (8 December 2013). "Evidence for biogenic graphite in early Archaean Isua metasedimentary rocks". Nature Geoscience. doi:10.1038/ngeo2025.
31. ^ Jump up to:a b Borenstein, Seth (19 October 2015). "Hints of life on what was thought to be desolate early Earth". Excite. Yonkers, NY: Mindspark Interactive Network. Associated Press. Retrieved 2015-10-20.
32. Bell, Elizabeth A.; Boehnike, Patrick; Harrison, T. Mark; et al. (19 October 2015). "Potentially biogenic carbon preserved in a 4.1 billion-year-old zircon" (PDF). Proc. Natl. Acad. Sci. U.S.A. Washington, D.C.: National Academy of Sciences. 112: 14518–21. doi:10.1073/pnas.1517557112. ISSN 1091-6490. PMC 4664351 . PMID 26483481. Retrieved 2015-10-20. Early edition, published online before print.
33. "The Cambrian Period". University of California Museum of Paleontology. Retrieved May 17, 2012.
34. Sahney, S.; Benton, M.J. & Falcon-Lang, H.J. (2010). "Rainforest collapse triggered Pennsylvanian tetrapod diversification in Euramerica" (PDF). Geology.

38 (12): 1079–1082. Bibcode:2010Geo....38.1079S. doi:10.1130/G31182.1.

35. Sahney, S. & Benton, M.J. (2008). "Recovery from the most profound mass extinction of all time" (PDF). Proceedings of the Royal Society: Biological. 275 (1636): 759–65. doi:10.1098/rspb.2007.1370. PMC 2596898 . PMID 18198148.
36. Bambach, R.K.; Knoll, A.H.; Wang, S.C. (December 2004). "Origination, extinction, and mass depletions of marine diversity". Paleobiology. 30 (4): 522–42. doi:10.1666/0094-8373(2004)030<0522:OEAMDO>2.0.CO;2. ISSN 0094-8373. Retrieved 2008-01-24.
37. Sala, Osvaldo E.; Meyerson, Laura A.; Parmesan, Camille (26 January 2009). Biodiversity change and human health: from ecosystem services to spread of disease. Island Press. pp. 3–5. ISBN 978-1-59726-497-6. Retrieved 28 June 2011.
38. Dasmann, Raymond Fredric (1968). A Different Kind of Country. Collier Books. ISBN 978-0-02-072810-8.
39. Soulé, Michael E.; Wilcox, Bruce A. (1980). Conservation biology: an evolutionary-ecological perspective. Sunderland, Mass: Sinauer Associates. ISBN 0-87893-800-1.
40. "Robert E. Jenkins". Nature.org. 2011-08-18. Retrieved 2011-09-24.
41. Wilson, E.O. (1 January 1988). Biodiversity. National Academies Press. ISBN 978-0-309-03739-6. online edition
42. Global Biodiversity Assessment: Summary for Policy-makers. Cambridge University Press. 1995. ISBN 978-0-521-56481-6. Annex 6, Glossary. Used as source by "Biodiversity", Glossary of terms related to the CBD, Belgian Clearing-House Mechanism. Retrieved 2006-04-26.
43. http://www.mass.gov/eea/agencies/dfg/dfw/natural-heritage/
44. Tor-Björn Larsson (2001). Biodiversity evaluation tools for European forests. Wiley-Blackwell. p. 178. ISBN 978-87-16-16434-6. Retrieved 28 June 2011.
45. Davis. Intro To Env Engg (Sie), 4E. McGraw-Hill Education (India) Pvt Ltd. pp. 4–. ISBN 978-0-07-067117-1. Retrieved 28 June 2011.
46. Lefcheck, Jon. "What is functional diversity, and why do we care?". sample(ECOLOGY). Retrieved 2015-12-22.
47. Campbell, AK (2003). "Save those molecules: molecular biodiversity and life". Journal of Applied Ecology. 40 (2): 193–203. doi:10.1046/j.1365-2664.2003.00803.x.
48. Wilcox, Bruce A. 1984. In situ conservation of genetic resources: determinants of minimum area requirements. In National Parks, Conservation and Development, Proceedings of the World Congress on National Parks,, J.A. McNeely and K.R. Miller, Smithsonian Institution Press, pp. 18–30.
49. D. L. Hawksworth (1996). Biodiversity: measurement and estimation. Springer. p. 6. ISBN 978-0-412-75220-9. Retrieved 28 June 2011.
50. Gaston, Kevin J.; Spicer, John I. (13 February 2004). Biodiversity: An Introduction. Wiley. ISBN 978-1-4051-1857-6.

51. Sahney, S.; Benton, M.J.; Ferry, Paul (2010). "Links between global taxonomic diversity, ecological diversity and the expansion of vertebrates on land". Biology Letters. The Royal Society. 6 (4): 544–7. doi:10.1098/rsbl.2009.1024. PMC 2936204 . PMID 20106856.

52. "A Durable Yet Vulnerable Eden in Amazonia". Dot Earth blog, New York Times. 2010-01-20. Retrieved 2013-02-02.

53. Margot S. Bass; Matt Finer; Clinton N. Jenkins; Holger Kreft; Diego F. Cisneros-Heredia; Shawn F. McCracken; Nigel C. A. Pitman; Peter H. English; Kelly Swing; Gorky Villa; Anthony Di Fiore; Christian C. Voigt; Thomas H. Kunz (2010). "Global Conservation Significance of Ecuador's Yasuní National Park". Public Library of Science. 5 (1): e8767. Bibcode:2010PLoSO...5.8767B. doi:10.1371/journal.pone.0008767. Retrieved 2011-06-07.

54. Benton M. J. (2001). "Biodiversity on land and in the sea". Geological Journal. 36 (3–4): 211–230. doi:10.1002/gj.877.

55. Mora, C.; et al. (2011). "How Many Species Are There on Earth and in the Ocean?". PLoS Biology. 9 (8): e1001127. doi:10.1371/journal.pbio.1001127. PMC 3160336 . PMID 21886479.

56. Mora C & Robertson DR (2005). "Causes of latitudinal gradients in species richness: a test with fishes of the Tropical Eastern Pacific"(PDF). Ecology. 86 (7): 1771–1792. doi:10.1890/04-0883.

57. Currie, D. J.; Mittelbach, G. G.; Cornell, H. V.; Kaufman, D. M.; Kerr, J. T.; Oberdorff, T. (2004). "A critical review of species-energy theory". Ecology Letters. 7: 1121–1134. doi:10.1111/j.1461-0248.2004.00671.x.

58. Allen A. P.; Gillooly J. F.; Savage V. M.; Brown J. H. (2006). "Kinetic effects of temperature on rates of genetic divergence and speciation". PNAS. 103 (24): 9130–9135. Bibcode:2006PNAS..103.9130A. doi:10.1073/pnas.0603587103. PMC 1474011 . PMID 16754845.

59. Hillebrand H (2004). "On the generality of the latitudinal diversity gradient". The American Naturalist. 163 (2): 192–211. doi:10.1086/381004. PMID 14970922.

60. "Moustakas, A. & I. Karakassis. How diverse is aquatic biodiversity research?, Aquatic Ecology, 39, 367–375" (PDF).

61. Morand, Serge; Krasnov, Boris R. (1 September 2010). The Biogeography of Host-Parasite Interactions. Oxford University Press. pp. 93–94. ISBN 978-0-19-956135-3. Retrieved 28 June 2011.

62. Biodiversity A-Z. "Biodiversity Hotspots".

63. Myers N (1988). "Threatened biotas: 'hot spots' in tropical forests". Environmentalist. 8 (3): 187–208. doi:10.1007/BF02240252. PMID 12322582.

64. Myers N (1990). "The biodiversity challenge: expanded hot-spots analysis". Environmentalist. 10 (4): 243–256. doi:10.1007/BF02239720. PMID 12322583.

65. Tittensor D.; et al. (2011). "Global patterns and predictors of marine biodiversity across taxa" (PDF). Nature. 466 (7310): 1098–1101. Bibcode:2010Natur.466.1098T. doi:10.1038/nature09329. PMID 20668450.

66. McKee, Jeffrey K. (December 2004). Sparing Nature: The Conflict Between Human Population Growth and Earth's Biodiversity. Rutgers University Press. p. 108. ISBN 978-0-8135-3558-6. Retrieved 28 June 2011.

67. "Colombia in the World". Alexander von Humboldt Institute for Research on Biological Resources. Retrieved 2013-12-30.

68. Normile, Dennis (10 September 2010). "Saving Forests to Save Biodiversity". Science. 329 (5997): 1278–1280. Bibcode:2010Sci...329.1278N. doi:10.1126/science.329.5997.1278. PMID 20829464. Retrieved 28 December 2010.

69. White, Gilbert (1887). "letter xx". The Natural History of Selborne: With A Naturalist's Calendar & Additional Observations. Scott.

70. Rosing, M.; Bird, D.; Sleep, N.; Bjerrum, C. (2010). "No climate paradox under the faint early Sun". Nature. 464 (7289): 744–747. Bibcode:2010Natur.464..744R. doi:10.1038/nature08955. PMID 20360739.

71. Alroy, J; Marshall, CR; Bambach, RK; Bezusko, K; Foote, M; Fursich, FT; Hansen, TA; Holland, SM; et al. (2001). "Effects of sampling standardization on estimates of Phanerozoic marine diversification". Proceedings of the National Academy of Sciences of the United States of America. 98 (11): 6261–6. Bibcode:2001PNAS...98.6261A. doi:10.1073/pnas.111144698. PMC 33456 . PMID 11353852.

72. "Mapping the web of life". Unep.org. Retrieved 2009-06-21.

73. Okasha, S. (2010). "Does diversity always grow?". Nature. 466 (7304): 318. Bibcode:2010Natur.466..318O. doi:10.1038/466318a.

74. Markov, AV; Korotaev, AV (2008). "Hyperbolic growth of marine and continental biodiversity through the phanerozoic and community evolution". Journal of General Biology. 69 (3): 175–94. PMID 18677962.

75. Markov, A; Korotayev, A (2007). "Phanerozoic marine biodiversity follows a hyperbolic trend". Palaeoworld. 16 (4): 311–318. doi:10.1016/j.palwor.2007.01.002.

76. National Survey Reveals Biodiversity Crisis American Museum of Natural History

77. Wilson, Edward O. (1 January 2002). The Future of Life. Alfred A. Knopf. ISBN 978-0-679-45078-8.

78. Cardinale, Bradley; et al. (2012). "Biodiversity loss and its impact on humanity". Nature. 486 (7401): 59–67. Bibcode:2012Natur.486...59C. doi:10.1038/nature11148. PMID 22678280.

79. Daniel, T. C.; et al. (21 May 2012). "Contributions of cultural services to the ecosystem services agenda". Proceedings of the National Academy of Sciences. 109 (23): 8812–8819. Bibcode:2012PNAS..109.8812D. doi:10.1073/pnas.1114773109.

80. Cardinale, Bradley. J.; et al. (March 2011). "The functional role of producer diversity in ecosystems". American Journal of Botany. 98 (3): 572–592. doi:10.3732/ajb.1000364. PMID 21613148.

81. Kiaer, Lars P.; Skovgaard, M.; Østergård, Hanne (1 December 2009). Grain yield increase in cereal variety mixtures: A meta-analysis of field trials. Field

Crops Research. 114 (3): 361–373. doi:10.1016/j.fcr.2009.09.006.

82. Letourneau, Deborah K. (1 January 2011). Does plant diversity benefit agroecosystems? A synthetic review. Ecological Applications. 21 (1): 9–21. doi:10.1890/09-2026.1.

83. Piotto, Daniel (1 March 2008). A meta-analysis comparing tree growth in monocultures and mixed plantations. Forest Ecology and Management. 255 (3–4): 781–786. doi:10.1016/j.foreco.2007.09.065.

84. Futuyma, Douglas J.; Shaffer, H. Bradley; Simberloff, Daniel, eds. (1 January 2009). Annual Reviews of Ecology, Evolution and Systematics: Vol 40 2009. Palo Alto, Calif.: Annual Reviews. pp. 573–592. ISBN 978-0-8243-1440-8.

85. Philpott, Stacy M.; Soong, Oliver; Lowenstein, Jacob H.; Pulido, Astrid Luz; Lopez, Diego Tobar (1 October 2009). Flynn, Dan F. B.; DeClerck, Fabrice. Functional richness and ecosystem services: bird predation on arthropods in tropical agroecosystems. Ecological Applications. 19 (7): 1858–1867. doi:10.1890/08-1928.1.

86. Van Bael, Sunshine A; et al. (Apr 2008). Birds as predators in tropical agroforestry systems. Ecology. 89 (4): 928–934. doi:10.1890/06-1976.1.

87. Vance-Chalcraft, Heather D.; et al. (1 November 2007). The Influence of Intraguild Predation on Prey Suppression and Prey Release: A Meta-analysis. Ecology. 88 (11): 2689–2696. doi:10.1890/06-1869.1.

88. Quijas, Sandra; Schmid, Bernhard; Balvanera, Patricia (1 November 2010). Plant diversity enhances provision of ecosystem services: A new synthesis. Basic and Applied Ecology. 11 (7): 582–593. doi:10.1016/j.baae.2010.06.009.

89. Levine, Jonathan M.; Adler, Peter B.; Yelenik, Stephanie G. (6 September 2004). A meta-analysis of biotic resistance to exotic plant invasions. Ecology Letters. 7 (10): 975–989. doi:10.1111/j.1461-0248.2004.00657.x.

90. Crowder, David W.; et al. Organic agriculture promotes evenness and natural pest control. Nature. 466 (7302): 109–112. Bibcode:2010Natur.466..109C. doi:10.1038/nature09183.

91. Andow, D A (1 January 1991). Vegetational Diversity and Arthropod Population Response. Annual Review of Entomology. 36 (1): 561–586. doi:10.1146/annurev.en.36.010191.003021.

92. Keesing, Felicia; et al. (Dec 2010). Impacts of biodiversity on the emergence and transmission of infectious diseases. Nature. 468 (7324): 647–652. Bibcode:2010Natur.468..647K. doi:10.1038/nature09575. PMID 21124449.

93. Johnson, Pieter T. J.; et al. (13 February 2013). Biodiversity decreases disease through predictable changes in host community competence. Nature. 494 (7436): 230–233. Bibcode:2013Natur.494..230J. doi:10.1038/nature11883.

94. Garibaldi, L. A.; et al. (28 February 2013). Wild Pollinators Enhance Fruit Set of Crops Regardless of Honey Bee Abundance. Science. 339 (6127): 1608–1611. Bibcode:2013Sci...339.1608G. doi:10.1126/science.1230200.

95. Costanza, Robert; et al. (1997). The value of the world's ecosystem services

and natural capital. Nature. 387 (6630): 253–260. Bibcode:1997Natur.387..253C. doi:10.1038/387253a0.

96. Hassan, Rashid M.; et al. (2006). Ecosystems and human well-being: current state and trends : findings of the Condition and Trends Working Group of the Millennium Ecosystem Assessment. Island Press. p. 105. ISBN 978-1-55963-228-7.
97. Vandermeer, John H. (2011). The Ecology of Agroecosystems. Jones & Bartlett Learning. ISBN 978-0-7637-7153-9.
98. Rice Grassy Stunt Virus. Lumrix.net. Retrieved 2009-06-21.
99. Wahl, GM; Robert de Saint Vincent B; Derose, ML (1984). Effect of chromosomal position on amplification of transfected genes in animal cells. Nature. 307 (5951): 516–20. Bibcode:1984Natur.307..516W. doi:10.1038/307516a0. PMID 6694743.
100. Southern Corn Leaf Blight. Retrieved 2007-11-13.
101. World Health Organization and Secretariat of the Convention on Biological Diversity (2015) Connecting Global Priorities: Biodiversity and Human Health, a State of Knowledge Review . See also Website of the Secretariat of the Convention on Biological Diversity on biodiversity and health. Other relevant resources include Reports of the 1st and 2nd International Conferences on Health and Biodiversity. See also: Website of the UN COHAB Initiative
102. Chivian, Eric, ed. (15 May 2008). Sustaining Life: How Human Health Depends on Biodiversity. OUP USA. ISBN 978-0-19-517509-7.
103. Corvalán, Carlos; Hales, Simon (2005). Ecosystems and Human Well-being: Health Synthesis. World Health Organization. pp. 28–. ISBN 978-92-4-156309-3. |first3= missing |last3= in Authors list (help)
104. (2009) Climate Change and Biological Diversity Convention on Biological Diversity Retrieved November 5, 2009
105. Ramanujan, Krishna (2 December 2010). Study: Loss of species is bad for your health. Cornell Chronicle. Retrieved 20 July 2011.
106. The World Bank (30 June 2010). Water and Development: An Evaluation of World Bank Support, 1997-2007. World Bank Publications. pp. 79–. ISBN 978-0-8213-8394-0.
107. Population Bulletin. Vol.63., No.3., p.8.
108. Gaston, Kevin J.; Warren, Philip H.; Devine-Wright, Patrick; Irvine, Katherine N.; Fuller, Richard A. (2007). Psychological benefits of greenspace increase with biodiversity. Biology Letters. 3 (4): 390–394. doi:10.1098/rsbl.2007.0149. PMC 2390667 . PMID 17504734.
109. COHAB Initiative: Biodiversity and Human Health - the issues. Cohabnet.org. Retrieved 2009-06-21.
110. World Wildlife Fund (WWF): Arguments for Protection website. Wwf.panda.org. Retrieved 2011-09-24.
111. Mendelsohn, Robert; Balick, Michael J. (1 April 1995). The value of undiscovered pharmaceuticals in tropical forests. Economic Botany. 49 (2): 223–228.

doi:10.1007/BF02862929.

112. (2006) Molecular Pharming GMO Compass Retrieved November 5, 2009, GMOcompass.org

113. Mendelsohn, Robert; Balick, Michael J. (1 July 1997). Notes on economic plants. Economic Botany. 51 (3): 328–328. doi:10.1007/BF02862103.

114. Harvey, Alan L. (2008-10-01). Natural products in drug discovery. Drug Discovery Today. 13 (19–20): 894–901. doi:10.1016/j.drudis.2008.07.004.

115. Hawkins E.S., Reich; Reich, MR (1992). Japanese-originated pharmaceutical products in the United States from 1960 to 1989: an assessment of innovation. Clin Pharmacol Ther. 51 (1): 1–11. doi:10.1038/clpt.1992.1. PMID 1732073.

116. Roopesh, J.; et al. (10 February 2008). Marine organisms: Potential Source for Drug Discovery (PDF). Current Science. 94 (3): 292.

117. Dhillion, SS; Svarstad, H; Amundsen, C; Bugge, HC (2002). Bioprospecting: Effects on environment and development. Ambio. 31 (6): 491–3. doi:10.1639/0044-7447(2002)031[0491:beoead]2.0.co;2. JSTOR 4315292. PMID 12436849.

118. Cole, A. (2005-07-16). Looking for new compounds in sea is endangering ecosystem. BMJ. 330 (7504): 1350. doi:10.1136/bmj.330.7504.1350-d.

119. COHAB Initiative - on Natural Products and Medicinal Resources. Cohabnet.org. Retrieved 2009-06-21.

120. IUCN, WRI, World Business Council for Sustainable Development, Earthwatch Inst. 2007 Business and Ecosystems: Ecosystem Challenges and Business Implications

121. Millennium Ecosystem Assessment 2005 Ecosystems and Human Well-being: Opportunities and Challenges for Business and Industry

122. Business and Biodiversity webpage of the U.N. Convention on Biological Diversity. Cbd.int. Retrieved 2009-06-21.

123. WRI Corporate Ecosystem Services Review. See also: Examples of Ecosystem-Service Based Risks, Opportunities and Strategies

124. Corporate Biodiversity Accounting. See also: Making the Natural Capital Declaration Accountable.

125. Tribot, A.,Mouquet, N., Villeger, S., Raymond, M., Hoff, F., Boissery, P., Holon, F., Deter, J. (2016). Taxonomic and functional diversity increase the aesthetic value of coralligenous reefs Scientific Reports, DOI: 10.1038/srep34229

126. Broad, William (November 19, 1996). Paradise Lost: Biosphere Retooled as Atmospheric Nightmare. The New York Times. Retrieved 10 April 2013.

127. LOSEY, JOHN E.; VAUGHAN, MACE (1 January 2006). The Economic Value of Ecological Services Provided by Insects. BioScience. 56 (4): 311. doi:10.1641/0006-3568(2006)56[311:TEVOES]2.0.CO;2.

128. Mora, Camilo; Tittensor, Derek P.; Adl, Sina; Simpson, Alastair G. B.; Worm, Boris; Mace, Georgina M. (23 August 2011). How Many Species Are There on Earth and in the Ocean?. PLoS Biology. 9 (8): e1001127. doi:10.1371/journal.

pbio.1001127. PMC 3160336 . PMID 21886479.

129. Wilson, J. Bastow; Peet, Robert K.; Dengler, Jürgen; Pärtel, Meelis (1 August 2012). Plant species richness: the world records. Journal of Vegetation Science. 23 (4): 796–802. doi:10.1111/j.1654-1103.2012.01400.x.

130. Appeltans, Ward; Ahyong, Shane T.; Anderson, Gary; Angel, Martin V.; Artois, Tom (1 December 2012). Bailly, Nicolas; Bamber, Roger; Barber, Anthony; Bartsch, Ilse; Berta, Annalisa; Błażewicz-Paszkowycz, Magdalena; Bock, Phil; Boxshall, Geoff; Boyko, Christopher B.; Brandão, Simone Nunes; Bray, Rod A.; Bruce, Niel L.; Cairns, Stephen D.; Chan, Tin-Yam; Cheng, Lanna; Collins, Allen G.; Cribb, Thomas; Curini-Galletti, Marco; Dahdouh-Guebas, Farid; Davie, Peter J.F.; Dawson, Michael N.; De Clerck, Olivier; Decock, Wim; De Grave, Sammy; de Voogd, Nicole J.; Domning, Daryl P.; Emig, Christian C.; Erséus, Christer; Eschmeyer, William; Fauchald, Kristian; Fautin, Daphne G.; Feist, Stephen W.; Fransen, Charles H.J.M.; Furuya, Hidetaka; Garcia-Alvarez, Oscar; Gerken, Sarah; Gibson, David; Gittenberger, Arjan; Gofas, Serge; Gómez-Daglio, Liza; Gordon, Dennis P.; Guiry, Michael D.; Hernandez, Francisco; Hoeksema, Bert W.; Hopcroft, Russell R.; Jaume, Damià; Kirk, Paul; Koedam, Nico; Koenemann, Stefan; Kolb, Jürgen B.; Kristensen, Reinhardt M.; Kroh, Andreas; Lambert, Gretchen; Lazarus, David B.; Lemaitre, Rafael; Longshaw, Matt; Lowry, Jim; Macpherson, Enrique; Madin, Laurence P.; Mah, Christopher; Mapstone, Gill; McLaughlin, Patsy A.; Mees, Jan; Meland, Kenneth; Messing, Charles G.; Mills, Claudia E.; Molodtsova, Tina N.; Mooi, Rich; Neuhaus, Birger; Ng, Peter K.L.; Nielsen, Claus; Norenburg, Jon; Opresko, Dennis M.; Osawa, Masayuki; Paulay, Gustav; Perrin, William; Pilger, John F.; Poore, Gary C.B.; Pugh, Phil; Read, Geoffrey B.; Reimer, James D.; Rius, Marc; Rocha, Rosana M.; Saiz-Salinas, José I.; Scarabino, Victor; Schierwater, Bernd; Schmidt-Rhaesa, Andreas; Schnabel, Kareen E.; Schotte, Marilyn; Schuchert, Peter; Schwabe, Enrico; Segers, Hendrik; Self-Sullivan, Caryn; Shenkar, Noa; Siegel, Volker; Sterrer, Wolfgang; Stöhr, Sabine; Swalla, Billie; Tasker, Mark L.; Thuesen, Erik V.; Timm, Tarmo; Todaro, M. Antonio; Turon, Xavier; Tyler, Seth; Uetz, Peter; van der Land, Jacob; Vanhoorne, Bart; van Ofwegen, Leen P.; van Soest, Rob W.M.; Vanaverbeke, Jan; Walker-Smith, Genefor; Walter, T. Chad; Warren, Alan; Williams, Gary C.; Wilson, Simon P.; Costello, Mark J.. The Magnitude of Global Marine Species Diversity. Current Biology. 22 (23): 2189–2202. doi:10.1016/j.cub.2012.09.036.

131. Encyclopedia Smithsonian: Numbers of Insects. Si.edu. Retrieved 2009-06-21.

132. Le Monde newspaper article (in French)

133. Proceedings of the National Academy of Sciences, Census of Marine Life (CoML) News.BBC.co.uk

134. Hawksworth, D. L. (24 July 2012). Global species numbers of fungi: are tropical studies and molecular approaches contributing to a more robust estimate?. Biodiversity and Conservation. 21 (9): 2425–2433. doi:10.1007/s10531-012-0335-x.

135. Hawksworth, D (2001). The magnitude of fungal diversity: The 1.5 million species

estimate revisited. Mycological Research. 105 (12): 1422–1432. doi:10.1017/S0953756201004725.

136. Acari at University of Michigan Museum of Zoology Web Page. Insects.ummz.lsa.umich.edu. 2003-11-10. Retrieved 2009-06-21.
137. Fact Sheet - Expedition Overview (PDF). J. Craig Venter Institute. Retrieved 29 August 2010.
138. Mirsky, Steve (March 21, 2007). Naturally Speaking: Finding Nature's Treasure Trove with the Global Ocean Sampling Expedition. Scientific American. Retrieved 4 May 2011.
139. Article collections published by the Public Library of Science. PLoS Collections. Retrieved 2011-09-24.
140. McKie, Robin (2005-09-25). Discovery of new species and extermination at high rate. The Guardian. London.
141. Bautista, L.M.; Pantoja, J.C. (2005). What species should we study next? (PDF). Bull. British Ecol. Soc. 36 (4): 27–28.
142. Gabriel, Sigmar (2007-03-09). 30% of all species lost by 2050. BBC News.
143. Reid Reversing loss of Biodiversity. Ag.arizona.edu. Retrieved 2009-06-21.
144. Pimm, S. L.; Russell, G. J.; Gittleman, J. L.; Brooks, T. M. (1995). The Future of Biodiversity (PDF). Science. 269 (5222): 347–350. Bibcode:1995Sci...269..347P. doi:10.1126/science.269.5222.347. PMID 17841251.
145. Researches find threat from biodiversity loss equals climate change threat. Winnipeg Free Press. 2012-06-07.
146. Living Planet Report 2014 (PDF), World Wildlife Fund, retrieved October 4, 2014
147. Endangered Species List Expands to 16,000. Retrieved 2007-11-13.
148. Moulton, Michael P.; Sanderson, James (1 September 1998). Wildlife Issues in a Changing World. CRC-Press. ISBN 978-1-56670-351-2.
149. Chen, Jim (2003). Across the Apocalypse on Horseback: Imperfect Legal Responses to Biodiversity Loss. The Jurisdynamics of Environmental Protection: Change and the Pragmatic Voice in Environmental Law. Environmental Law Institute. p. 197. ISBN 978-1-58576-071-8.
150. Hippo dilemma. Windows on the Wild. New Africa Books. 2005. ISBN 978-1-86928-380-3.
151. IUCN's Classification of Direct Threats. Conservationmeasures.org. Retrieved 2011-09-24.
152. Ehrlich, Paul R.; Ehrlich, Anne H. (1983). Extinction: The Causes and Consequences of the Disappearance of Species. Ballantine Books. ISBN 978-0-345-33094-9.
153. C.Michael Hogan. 2010. Deforestation Encyclopedia of Earth. ed. C.Cleveland. NCSE. Washington DC
154. Drakare, Stina; Lennon, Jack J.; Hillebrand, Helmut (2006). The imprint of the geographical, evolutionary and ecological context on species-area relationships.

Ecology Letters. 9 (2): 215–227. doi:10.1111/j.1461-0248.2005.00848.x. PMID 16958886.

155. Study: Loss Of Genetic Diversity Threatens Species Diversity. Enn.com. 2007-09-26. Retrieved 2009-06-21.

156. Science Connection 22 (July 2008)

157. Koh L. P.; Dunn R. R.; Sodhi N. S.; Colwell R. K.; Proctor H. C.; Smith V. S. (2004). Species Coextinctions and the Biodiversity Crisis(PDF). Science. 305 (5690): 1632–4. Bibcode:2004Sci...305.1632K. doi:10.1126/science.1101101. PMID 15361627. Archived from the original (PDF) on 2009-03-26.

158. Torchin, Mark E.; Lafferty, Kevin D.; Dobson, Andrew P.; McKenzie, Valerie J.; Kuris, Armand M. (6 February 2003). Introduced species and their missing parasites. Nature. 421 (6923): 628–630. Bibcode:2003Natur.421..628T. doi:10.1038/nature01346.

159. Levine, Jonathan M.; D'Antonio, Carla M. (1 October 1999). Elton Revisited: A Review of Evidence Linking Diversity and Invasibility. Oikos. 87 (1): 15. doi:10.2307/3546992.

160. Levine, J. M. (5 May 2000). Species Diversity and Biological Invasions: Relating Local Process to Community Pattern. Science. 288 (5467): 852–854. Bibcode:2000Sci...288..852L. doi:10.1126/science.288.5467.852. PMID 10797006.

161. GUREVITCH, J; PADILLA, D (1 September 2004). Are invasive species a major cause of extinctions?. Trends in Ecology & Evolution. 19 (9): 470–474. doi:10.1016/j.tree.2004.07.005.

162. Sax, Dov F.; Gaines, Steven D.; Brown, James H. (1 December 2002). Species Invasions Exceed Extinctions on Islands Worldwide: A Comparative Study of Plants and Birds. The American Naturalist. 160 (6): 766–783. doi:10.1086/343877. PMID 18707464.

163. Jude, David auth., ed. by M. Munawar (1995). The lake Huron ecosystem: ecology, fisheries and management. Amsterdam: S.P.B. Academic Publishing. ISBN 90-5103-117-3.

164. Are invasive plants a threat to native biodiversity? It depends on the spatial scale. ScienceDaily. 11 April 2011.

165. Mooney, H. A.; Cleland, EE (2001). The evolutionary impact of invasive species. Proceedings of the National Academy of Sciences. 98 (10): 5446–5451. Bibcode:2001PNAS...98.5446M. doi:10.1073/pnas.091093398. PMC 33232 . PMID 11344292.

166. Glossary: definitions from the following publication: Aubry, C., R. Shoal and V. Erickson. 2005. Grass cultivars: their origins, development, and use on national forests and grasslands in the Pacific Northwest. USDA Forest Service. 44 pages, plus appendices.; Native Seed Network (NSN), Institute for Applied Ecology, 563 SW Jefferson Ave, Corvallis, OR 97333, USA. Nativeseednetwork.org. Retrieved 2009-06-21.

167. Rhymer, Judith M.; Simberloff, Daniel (1996). Extinction by Hybridization

and Introgression. Annual Review of Ecology and Systematics. 27: 83–109. doi:10.1146/annurev.ecolsys.27.1.83. JSTOR 2097230.

168. Potts, Bradley M.; Barbour, Robert C.; Hingston, Andrew B. (2001). Genetic Pollution from Farm Forestry Using Eucalypt Species and Hydrids: A Report for the RIRDC/L & WA/FWPRDC Joint Venture Agroforestry Program. RIRDC. ISBN 978-0-642-58336-9. ISSN 1440-6845. RIRDC.gov.au RIRDC Publication No 01/114; RIRDC Project No CPF - 3A; Australian Government, Rural Industrial Research and Development Corporation Archived 5 January 2016 at the Wayback Machine.

169. Grafton, R. Q.; Kompas, T.; Hilborn, R. W. (2007). Economics of Overexploitation Revisited. Science. 318 (5856): 1601–1601. Bibcode:2007Sci...318.1601G. doi:10.1126/science.1146017.

170. Burney, D. A.; Flannery, T. F. (July 2005). Fifty millennia of catastrophic extinctions after human contact (PDF). Trends in Ecology & Evolution. Elsevier. 20 (7): 395–401. doi:10.1016/j.tree.2005.04.022. PMID 16701402. Archived from the original (PDF) on 10 June 2010. Retrieved 2009-06-12.

171. Genetic Pollution: The Great Genetic Scandal; Archived 18 May 2009 at the Wayback Machine.

172. Pollan, Michael (2001-12-09). The year in ideas: A TO Z.; Genetic Pollution; By Michael Pollan, The New York Times, December 9, 2001. New York Times. Retrieved 2009-06-21.

173. Ellstrand, Norman C. (2003). Dangerous Liaisons? When Cultivated Plants Mate with Their Wild Relatives. The Johns Hopkins University Press. ISBN 0-8018-7405-X. Reviewed in Strauss, Steven H; DiFazio, Stephen P (2004-01-01). Hybrids abounding. Nature Biotechnology. Nature.com. 22 (1): 29–30. doi:10.1038/nbt0104-29.

174. Zaid, A. (1999). Genetic pollution: Uncontrolled spread of genetic information. Glossary of Biotechnology and Genetic Engineering. Food and Agriculture Organization of the United Nations. ISBN 978-92-5-104369-1. Retrieved 2009-06-21.

175. Genetic pollution: Uncontrolled escape of genetic information (frequently referring to products of genetic engineering) into the genomes of organisms in the environment where those genes never existed before. Searchable Biotechnology Dictionary, University of Minnesota, Boku.ac.at Archived 10 February 2008 at the Wayback Machine.

176. The many facets of pollution. Bologna University. Retrieved May 18, 2012.

177. Climate change and biodiversity (PDF). Intergovernmental Panel on Climate Change. 2005.

178. Kannan, R.; James, D. A. (2009). Effects of climate change on global biodiversity: a review of key literature (PDF). Tropical Ecology. 50 (1): 31–39. ISSN 0564-3295. Retrieved 2014-05-21.

179. Climate change, reefs and the Coral Triangle. wwf.panda.org. Retrieved 2015-11-09.

180. Aldred, Jessica. Caribbean coral reefs ‹will be lost within 20 years' without protection. the Guardian. Retrieved 2015-11-09.

181. Ainsworth, Elizabeth A.; Long, Stephen P. (18 November 2004). What have we learned from 15 years of free-air CO2 enrichment (FACE)? A meta-analytic review of the responses of photosynthesis, canopy properties and plant production to rising CO2. New Phytologist. 165 (2): 351–372. doi:10.1111/j.1469-8137.2004.01224.x. PMID 15720649.

182. Doney, Scott C.; Fabry, Victoria J.; Feely, Richard A.; Kleypas, Joan A. (1 January 2009). Ocean Acidification: The Other CO Problem. Annual Review of Marine Science. 1 (1): 169–192. Bibcode:2009ARMS....1..169D. doi:10.1146/annurev.marine.010908.163834.

183. Loarie, Scott R.; Duffy, Philip B.; Hamilton, Healy; Asner, Gregory P.; Field, Christopher B.; Ackerly, David D. (24 December 2009). The velocity of climate change. Nature. 462 (7276): 1052–1055. Bibcode:2009Natur.462.1052L. doi:10.1038/nature08649. PMID 20033047.

184. Walther, Gian-Reto; Roques, Alain; Hulme, Philip E.; Sykes, Martin T.; Pyšek, Petr (1 December 2009). Kühn, Ingolf; Zobel, Martin; Bacher, Sven; Botta-Dukát, Zoltán; Bugmann, Harald. Alien species in a warmer world: risks and opportunities. Trends in Ecology & Evolution. 24 (12): 686–693. doi:10.1016/j.tree.2009.06.008.

185. Lovejoy, Thomas E.; Hannah, Lee Jay (2005). Climate Change and Biodiversity. New Haven: Yale University Press. pp. 41–55. ISBN 978-0-300-10425-7.

186. Hegland, Stein Joar; Nielsen, Anders; Lázaro, Amparo; Bjerknes, Anne-Line; Totland, Ørjan (1 February 2009). How does climate warming affect plant-pollinator interactions?. Ecology Letters. 12 (2): 184–195. doi:10.1111/j.1461-0248.2008.01269.x.

187. Min, Seung-Ki; Xuebin Zhang; Francis W. Zwiers; Gabriele C. Hegerl (Feb 17, 2011). Human contribution to more-intense precipitation extremes. Nature. 470 (7334): 378–381. Bibcode:2011Natur.470..378M. doi:10.1038/nature09763.

188. Brown, Paul (2004-01-08). An unnatural disaster. The Guardian. London. Retrieved 2009-06-21.

189. Visconti, Piero; et. al (February 2015). Projecting global biodiversity indicators under future development scenarios. Conservation Letters. Wiley. doi:10.1111/conl.12159. Retrieved 2015-03-25.

190. World Population Growth, 1950–2050. Population Reference Bureau.

191. World population to keep growing this century, hit 11 billion by 2100. UWToday. September 18, 2014.

192. Harris, Paul (October 22, 2011). Population of world ‹could grow to 15bn by 2100'. The Guardian. Retrieved November 20, 2016.

193. Citizens arrest. The Guardian. July 11, 2007.

194. Population Bomb Author's Fix For Next Extinction: Educate Women. Scientific American. August 12, 2008.

195. Dumont, E. (2012). Estimated impact of global population growth on future wilderness extent. (PDF). Earth System Dynamics Discussions. 3: 433–452. Bibcode:2012ESDD....3..433D. doi:10.5194/esdd-3-433-2012.

196. Wake D. B.; Vredenburg V. T. (2008). Are we in the midst of the sixth mass extinction? A view from the world of amphibians (PDF). Proceedings of the National Academy of Sciences of the United States of America. 105: 11466–11473. Bibcode:2008PNAS..10511466W. doi:10.1073/pnas.0801921105. PMC 2556420 . PMID 18695221.

197. Koh, LP; Dunn, RR; Sodhi, NS; Colwell, RK; Proctor, HC; Smith, VS (2004). Species coextinctions and the biodiversity crisis (PDF). Science. 305 (5690): 1632–4. Bibcode:2004Sci...305.1632K. doi:10.1126/science.1101101. PMID 15361627.[dead link]

198. McCallum M. L. (2007). Amphibian Decline or Extinction? Current Declines Dwarf Background Extinction Rate (PDF). Journal of Herpetology. 41 (3): 483–491. doi:10.1670/0022-1511(2007)41[483:ADOECD]2.0.CO;2. ISSN 0022-1511.

199. Jackson, J. B. C. (2008). Colloquium Paper: Ecological extinction and evolution in the brave new ocean. Proceedings of the National Academy of Sciences. 105: 11458–11465. Bibcode:2008PNAS..10511458J. doi:10.1073/pnas.0802812105. PMC 2556419 . PMID 18695220.

200. Dunn R. R. (2005). Modern Insect Extinctions, the Neglected Majority (PDF). Conservation Biology. 19 (4): 1030–1036. doi:10.1111/j.1523-1739.2005.00078.x. Archived from the original(PDF) on 8 July 2009.

201. Ceballos, Gerardo; Ehrlich, Paul R.; Barnosky, Anthony D.; García, Andrés; Pringle, Robert M.; Palmer, Todd M. (2015). Accelerated modern human–induced species losses: Entering the sixth mass extinction. Science Advances. 1 (5): e1400253. doi:10.1126/sciadv.1400253.

202. Costanza, R.; d'Arge, R.; de Groot, R.; Farberk, S.; Grasso, M.; Hannon, B.; Limburg, Karin; Naeem, Shahid; et al. (1997). The value of the world's ecosystem services and natural capital (PDF). Nature. 387 (6630): 253–260. Bibcode:1997Natur.387..253C. doi:10.1038/387253a0.

203. Millennium Ecosystem Assessment (2005). World Resources Institute, Washington, DC. Ecosystems and Human Well-being: Biodiversity Synthesis

204. Soulé, Michael E. (1986). What is conservation biology?. BioScience. 35 (11): 727–734. doi:10.2307/1310054. JSTOR 1310054.

205. Davis, Peter (1996). Museums and the natural environment: the role of natural history museums in biological conservation. Leicester University Press. ISBN 978-0-7185-1548-5.

206. Dyke, Fred Van (29 February 2008). Conservation Biology: Foundations, Concepts, Applications. Springer Science & Business Media. ISBN 978-1-4020-6890-4.

207. Hunter, Malcolm L. (1996). Fundamentals of Conservation Biology. Blackwell Science. ISBN 978-0-86542-371-8.

208. Bowen, B. W. (1999). Preserving genes, species, or ecosystems? Healing the

fractured foundations of conservation policy. Molecular Ecology. 8: S5–S10. doi:10.1046/j.1365-294x.1999.00798.x.

209. Soulé, Michael E. (1 January 1986). Conservation Biology: The Science of Scarcity and Diversity. Sinauer Associates. ISBN 978-0-87893-794-3.

210. Margules C. R.; Pressey R. L. (2000). Systematic conservation planning (PDF). Nature. 405 (6783): 243–253. doi:10.1038/35012251. PMID 10821285.

211. Example: Gascon, C., Collins, J. P., Moore, R. D., Church, D. R., McKay, J. E. and Mendelson, J. R. III (eds) (2007). Amphibian Conservation Action Plan. IUCN/SSC Amphibian Specialist Group. Gland, Switzerland and Cambridge, UK. 64pp. Amphibians.org, see also Millenniumassessment.org, Europa.eu Archived 30 December 2008 at the Wayback Machine.

212. Luck, Gary W.; Daily, Gretchen C.; Ehrlich, Paul R. (2003). Population diversity and ecosystem services (PDF). Trends in Ecology & Evolution. 18 (7): 331–336. doi:10.1016/S0169-5347(03)00100-9. Archived from the original (PDF) on 19 February 2006.

213. Millenniumassessment.org Archived 13 August 2015 at the Wayback Machine.

214. Beantwoording vragen over fokken en doden van gezonde dieren in dierentuinen (PDF) (in Dutch). Ministry of Economic Affairs (Netherlands). 25 March 2014. Retrieved 9 June 2014.

215. Barcode of Life. Barcoding.si.edu. 2010-05-26. Retrieved 2011-09-24.

216. Eradication of exotic animals (camels) in Australia[dead link]

217. Belgium creating 45 seed gardens; gene banks with intent to reintroduction. Hbvl.be. 2011-09-08. Retrieved 2011-09-24.

218. Conservationists Use Triage to Determine which Species to Save and Not; Like battlefield medics, conservationists are being forced to explicitly apply triage to determine which creatures to save and which to let go July 23, 2012 Scientific American.

219. Jones-Walters, L.; Mulder, I. (2009). Valuing nature: The economics of biodiversity (PDF). Journal for Nature Conservation. 17 (4): 245–247. doi:10.1016/j.jnc.2009.06.001.

220. Gene Patenting. Ornl.gov. Retrieved 2009-06-21.

221. Fred Bosselman, A Dozen Biodiversity Puzzles, 12 N.Y.U. Environmental Law Journal 364 (2004) (PDF). Retrieved 2011-09-24.

222. Wilson Edward O (2000). On the Future of Conservation Biology. Conservation Biology. 14 (1): 1–3. doi:10.1046/j.1523-1739.2000.00000-e1.x.

223. Nee S (2004). More than meets the eye. Nature. 429 (6994): 804–805. Bibcode:2004Natur.429..804N. doi:10.1038/429804a. PMID 15215837.

224. Stork, Nigel E. (2007). Biodiversity: World of insects. Nature. 448 (7154): 657–658. Bibcode:2007Natur.448..657S. doi:10.1038/448657a. PMID 17687315.

225. Thomas J. A.; Telfer M. G.; Roy D. B.; Preston C. D.; Greenwood J. J. D.; Asher J.; Fox R.; Clarke R. T.; Lawton J. H. (2004). Comparative Losses of British

Butterflies, Birds, and Plants and the Global Extinction Crisis. Science. 303 (5665): 1879–1881. Bibcode:2004Sci...303.1879T. doi:10.1126/science.1095046. PMID 15031508.

226. Dunn, Robert R. (2005). Modern Insect Extinctions, the Neglected Majority. Conservation Biology. 19 (4): 1030–1036. doi:10.1111/j.1523-1739.2005.00078.x.

CHAPTER 2

Global Biodiversity

Chapter Contents

- **Introduction**
- **Extinction**
- **Speciation**
- **Background Extinction Rate**

Biodiversity (or biological diversity) is a collective term meaning the totality and variety of life on Earth. Biodiversity includes genetic diversity within species, the variety among species, and the range of ecosystems within which life exists and interacts.

INTRODUCTION

Estimates on the number of Earth's current species range from 2 million to 10^12, of which about 1.6 million have been databased thus far and over 80 percent have not yet been described. In May 2016, scientists reported that 1 trillion species are estimated to be on Earth currently with only one-thousandth of one percent described. The total amount of DNA base pairs on Earth, as a possible approximation of global biodiversity, is estimated at 5.0×10^{37}, and weighs 50 billion tonnes. In comparison, the total mass of the biosphere has been estimated to be as much as 4 TtC (trillion tons of carbon).

In other related studies, around 1.9 million extant species are believed to have been described currently, but some scientists believe 20% are synonyms, reducing the total valid described species to 1.5 million. In 2013, a study published in Science estimated there to be 5 ± 3 million extant species on Earth. Another study, published in 2011 by PLoS Biology, estimated there to be 8.7 million ± 1.3 million eukaryotic species on Earth. Some 250,000 valid fossil species have been described, but this is believed to be a small proportion of all species that have ever lived.

Global biodiversity is affected by extinction and speciation. The background extinction rate varies among taxa but it is estimated that there is approximately one extinction per million species years. Mammal species, for example, typically persist for 1 million years. Biodiversity has grown and shrunk in earth's past due to (presumably) abiotic factors such as extinction events caused by geologically rapid changes in climate. Climate change 299 million years ago was one such event. A cooling and drying resulted in catastrophic rainforest collapse and subsequently a great loss of diversity, especially of amphibians. However, the current rate and magnitude of extinctions are much higher than background estimates.This, considered by some to be leading to the sixth mass extinction, is a result of human impacts on the environment.

Examples of the biodiversity of the Earth.

EXTINCTION

In biology and ecology, extinction is the end of an organism or of a group of organisms (taxon), normally a species. The moment of extinction is generally considered to be the death of the last individual of the species, although the capacity to breed and recover may have been lost before this point. Because a species' potential range may be very large, determining this moment is difficult, and is usually done retrospectively. This difficulty leads to phenomena such as Lazarus taxa, where a species presumed extinct abruptly reappears (typically in the fossil record) after a period of apparent absence.

More than 99 percent of all species, amounting to over five billion species, that ever lived on Earth are estimated to be extinct. Estimates on the number of Earth's current species range from 10 million to 14 million, of which about 1.2 million have been documented and over 86 percent have not yet been described.

Through evolution, species arise through the process of speciation—where new varieties of organisms arise and thrive when they are able to find and exploit an ecological niche—and species become extinct when they are no longer able to survive in changing conditions or against superior competition. The relationship between animals and their ecological niches has been firmly established. A typical species becomes extinct within 10 million years of its first appearance, although some species, called living fossils, survive with virtually no morphological change for hundreds of millions of years.

Mass extinctions are relatively rare events; however, isolated extinctions are quite common. Only recently have extinctions been recorded and scientists have become alarmed at the current high rate of extinctions. Most species that become extinct are never scientifically documented. Some scientists estimate that up to half of presently existing plant and animal species may become extinct by 2100.

Skeleton of Palaeoloxodon namadicus, an extinct elephant species.

Skeleton of various extinct dinosaurs; some other dinosaur lineages still flourish in the form of birds.

The dodo of Mauritius, shown here in a 1626 illustration by Roelant Savery, is an often-cited example of modern extinction.

Definition

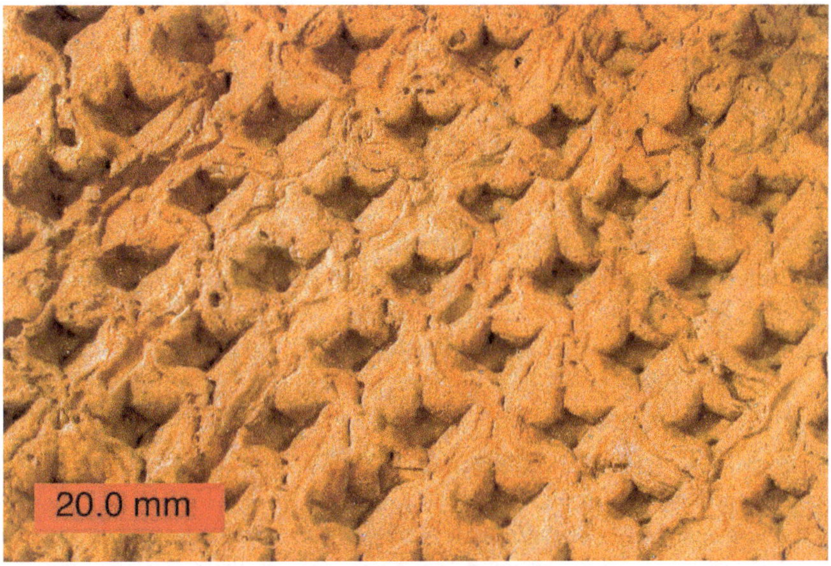

External mold of the extinct *Lepidodendron* from the Upper Carboniferous of Ohio.

A species is extinct when the last existing member dies. Extinction therefore becomes a certainty when there are no surviving individuals that can reproduce and create a new generation. A species may become functionally extinct when only a handful of individuals survive, which cannot reproduce due to poor health, age, sparse distribution over a large range, a lack of individuals of both sexes (in sexually reproducing species), or other reasons.

Pinpointing the extinction (or pseudoextinction) of a species requires a clear definition of that species. If it is to be declared extinct, the species in question must be uniquely distinguishable from any ancestor or daughter species, and from any other closely related species. Extinction of a species (or replacement by a daughter species) plays a key role in the punctuated equilibrium hypothesis of Stephen Jay Gould and Niles Eldredge.

In ecology, *extinction* is often used informally to refer to local extinction, in which a species ceases to exist in the chosen area of study, but may still exist elsewhere. This phenomenon is also known as extirpation. Local extinctions may be followed by a replacement of the species taken from other locations; wolf reintroduction is an example of this. Species which are not extinct are termed extant. Those that are extant but threatened by extinction are referred to as threatened or endangered species.

Currently an important aspect of extinction is human attempts to preserve critically endangered species. These are reflected by the creation of the conservation status extinct in the wild (EW). Species listed under this status by the International Union for Conservation of Nature (IUCN) are not known to have any living specimens in the wild, and are maintained only in zoos or other artificial environments. Some of these species are functionally extinct, as they are no longer part of their natural habitat and it is unlikely the species will ever be restored to the wild. When possible, modern zoological institutions try to maintain a viable population for species preservation and possible future reintroduction to the wild, through use of carefully planned breeding programs.

The extinction of one species' wild population can have knock-on effects, causing further extinctions. These are also called chains of extinction. This is especially common with extinction of keystone species.

Pseudoextinction

Descendants may or may not exist for extinct species. Daughter species that evolve from a parent species carry on most of the parent species' genetic information, and even though the parent species may become extinct, the daughter species lives on. Extinction of a parent species where daughter species or subspecies are still extant is called pseudoextinction. In other cases, species have produced no new variants, or none that are able to survive the parent species' extinction.

Pseudoextinction is difficult to demonstrate unless one has a strong chain of evidence linking a living species to members of a pre-existing species. For example, it is sometimes claimed that the extinct *Hyracotherium*, which was an early horse that shares a common ancestor with the modern horse, is pseudoextinct, rather than extinct, because there are several extant species of *Equus*, including zebra and donkey. However, as fossil species typically leave no genetic material behind, one cannot say whether *Hyracotherium* evolved into more modern horse species or merely evolved from a common ancestor with modern horses. Pseudoextinction is much easier to demonstrate for larger taxonomic groups.

Lazarus Taxa

The coelacanth, a fish related to lungfish and tetrapods, was considered to have been extinct since the end of the Cretaceous Period until 1938 when a specimen was found, off the Chalumna River (now Tyolomnqa) on the east coast of South Africa. Museum curator Marjorie Courtenay-Latimer discovered the fish among the catch of a local angler, Captain Hendrick Goosen, on December 23, 1938. A local chemistry professor, JLB Smith, confirmed the fish's importance with a famous cable: MOST IMPORTANT PRESERVE SKELETON AND GILLS = FISH DESCRIBED.

Far more recent possible or presumed extinctions of species which may turn out still to exist include the thylacine, or Tasmanian tiger (*Thylacinus cynocephalus*), the last known example of which died in Hobart Zoo in Tasmania in 1936; the Japanese wolf (*Canis lupus hodophilax*), last sighted over 100 years ago; the ivory-billed woodpecker (*Campephilus principalis*), last sighted for certain in 1944; and the slender-billed curlew (*Numenius tenuirostris*), not seen since 2007.

Causes

As long as species have been evolving, species have been going extinct. It is estimated that over 99.9% of all species that ever lived are extinct. The average lifespan of a species is 1-10 million years, although this varies widely between taxa. There are a variety of causes that can contribute directly or indirectly to the extinction of a species or group of species. "Just as each species is unique", writes Beverly and Stephen C. Stearns, "so is each extinction ... the causes for each are varied—some subtle and complex, others obvious and simple". Most simply, any species that cannot survive and reproduce in its environment and cannot move to a new environment where it can do so, dies out and becomes extinct. Extinction of a species may come suddenly when an otherwise healthy species is wiped out completely, as when toxic pollution renders its entire habitat unliveable; or may occur gradually over thousands or millions of years, such as when a species gradually loses out in competition for food to better adapted competitors. Extinction may occur a long time after the events that set it in motion, a phenomenon known as extinction debt.

The passenger pigeon, one of hundreds of species of extinct birds, was hunted to extinction over the course of a few decades.

Assessing the relative importance of genetic factors compared to environmental ones as the causes of extinction has been compared to the debate on nature and nurture. The question of whether more extinctions in the fossil record have been caused by evolution or by catastrophe is a subject of discussion. Mark Newman, the author of *Modeling Extinction*, argues for a mathematical model that falls between the two positions. By contrast, conservation biology uses the extinction vortex model to classify extinctions by cause. When concerns about human extinction have been raised, for example in Sir Martin Rees' 2003 book *Our Final Hour*, those concerns lie with the effects of climate change or technological disaster.

Currently, Environmental groups and some governments are concerned with the extinction of species caused by humanity, and they try to prevent further extinctions through a variety of conservation programs. Humans can cause extinction of a species through overharvesting, pollution, habitat destruction, introduction of invasive species (such as new predators and food competitors), overhunting, and other influences. Explosive, unsustainable human population growth is an essential cause of the extinction crisis. According to the International Union for Conservation of Nature (IUCN), 784 extinctions have been recorded since the year 1500, the arbitrary date selected to define recent extinctions, up to the year 2004; with many more likely to have gone unnoticed. Several species have also been listed as extinct since 2004.

Genetics and Demographic Phenomena

Population genetics and demographic phenomena affect the evolution, and therefore the risk of extinction of species. Limited geographic range is the most important determinant of genus extinction at background rates but becomes increasingly irrelevant as mass extinction arises.

Natural selection acts to propagate beneficial genetic traits and eliminate weaknesses. But a deleterious mutation can also be spread throughout a population through genetic drift.

Because traits are selected and not genes, the relationship between genetic diversity and extinction risk can be complex. Factors such as balancing selection, cryptic genetic variation, phenotypic plasticity, and degeneracy all potentially play roles.

A diverse or deep gene pool gives a population a higher chance of surviving an adverse change in conditions. Effects that cause or reward a loss in genetic diversity can increase the chances of extinction of a species. Population bottlenecks can dramatically reduce genetic diversity by severely limiting the number of reproducing individuals and make inbreeding more frequent. The founder effect can cause rapid, individual-based speciation and is the most dramatic example of a population bottleneck.

Genetic Pollution

Purebred wild species evolved to a specific ecology can be threatened with extinction through the process of genetic pollution—i.e., uncontrolled hybridization, introgression genetic swamping which leads to homogenization or out-competition from the introduced (or hybrid) species. Endemic populations can face such extinctions when new populations are imported or selectively bred by people, or when habitat modification brings previously isolated species into contact. Extinction is likeliest for rare species coming into contact with more abundant ones; interbreeding can swamp the rarer gene pool and create hybrids, depleting the purebred gene pool (for example, the endangered wild water buffalo is most threatened with extinction by genetic pollution from the abundant domestic water buffalo). Such extinctions are not always apparent from morphological (non-genetic) observations.

Some degree of gene flow is a normal evolutionarily process, nevertheless, hybridization (with or without introgression) threatens rare species' existence.

The gene pool of a species or a population is the variety of genetic information in its living members. A large gene pool (extensive genetic diversity) is associated with robust populations that can survive bouts of intense selection. Meanwhile, low genetic diversity (see inbreeding and population bottlenecks) reduces the range of adaptions possible. Replacing native with alien genes narrows genetic diversity within the original population, thereby increasing the chance of extinction.

Scorched land resulting from slash-and-burn agriculture.

Habitat Degradation

Habitat degradation is currently the main anthropogenic cause of species extinctions. The main cause of habitat degradation worldwide is agriculture, with urban sprawl, logging, mining, and some fishing practices close behind. The degradation of a species' habitat may alter the fitness landscape to such an extent that the species is no longer able to survive and becomes extinct. This may occur by direct effects, such as the environment becoming toxic, or indirectly, by limiting a species' ability to compete effectively for diminished resources or against new competitor species.

Habitat degradation through toxicity can kill off a species very rapidly, by killing all living members through contamination or sterilizing them. It can also occur over longer periods at lower toxicity levels by affecting life span, reproductive capacity, or competitiveness.

Habitat degradation can also take the form of a physical destruction of niche habitats. The widespread destruction of tropical rainforests and replacement with open pastureland is widely cited as an example of this; elimination of the dense forest eliminated the infrastructure needed by many species to survive. For example, a fern that depends on dense shade for protection from direct sunlight can no longer survive without forest to shelter it. Another example is the destruction of ocean floors by bottom trawling.

Diminished resources or introduction of new competitor species also often accompany habitat degradation. Global warming has allowed some species to expand their range, bringing unwelcome competition to other species that previously occupied that area. Sometimes these new competitors are predators and directly affect prey species, while at other times they may merely outcompete vulnerable species for limited resources. Vital resources including water and food can also be limited during habitat degradation, leading to extinction.

The golden toad was last seen on May 15, 1989. Decline in amphibian populations is ongoing worldwide.

Predation, Competition, and Disease

In the natural course of events, species become extinct for a number of reasons, including but not limited to: extinction of a necessary host, prey or pollinator, inter-species competition, inability to deal with evolving diseases and changing environmental conditions (particularly sudden changes) which can act to introduce novel predators, or to remove prey. Recently in geological time, humans have become an additional cause of extinction (many people would say premature extinction) of some species, either as a new mega-predator or by transporting animals and plants from one part of the world to another. Such introductions have been occurring for thousands of years, sometimes intentionally (e.g. livestock released by sailors on islands as a future source of food) and sometimes accidentally (e.g. rats escaping from boats). In most cases, the introductions are unsuccessful, but when an invasive alien species does become established, the consequences can be catastrophic. Invasive alien species can affect native species directly by eating them, competing with them, and introducing pathogens or parasites that sicken or kill them; or indirectly by destroying or degrading their habitat. Human populations may themselves act as invasive predators. According to the "overkill hypothesis", the swift extinction of the megafauna in areas such as Australia (40,000 years before present), North and South America (12,000 years before present), Madagascar, Hawaii (300–1000 CE), and New Zealand (1300–1500 CE), resulted from the sudden introduction of human beings to environments full of animals that had never seen them before, and were therefore completely unadapted to their predation techniques.

Coextinction

The large Haast's eagle and moa from New Zealand.

Coextinction refers to the loss of a species due to the extinction of another; for example, the extinction of parasitic insects following the loss of their hosts. Coextinction can also occur when a species loses its pollinator, or to predators in a food chain who lose their prey. "Species coextinction is a manifestation of the interconnectedness of organisms in complex ecosystems ... While coextinction may not be the most important cause of species extinctions, it is certainly an insidious one. Coextinction is especially common when a keystone species goes extinct. Models suggest that coextinction is the most common form of biodiversity loss. There may be a cascade of coextinction across the trophic levels. Such effects are most severe in mutualistic and parasitic relationships. An example of coextinction is the Haast's eagle and the moa: the Haast's eagle was a predator that became extinct because its food source became extinct. The moa were several species of flightless birds that were a food source for the Haast's eagle.

Climate Change

Extinction as a result of climate change has been confirmed by fossil studies. Particularly, the extinction of amphibians during the Carboniferous Rainforest Collapse, 305 million years ago. A 2003 review across 14 biodiversity research centers predicted that, because of climate change, 15–37% of land species would be "committed to extinction" by 2050. The ecologically rich areas that would potentially suffer the heaviest losses include the Cape Floristic Region, and the Caribbean Basin. These areas might see a doubling of present carbon dioxide levels and rising temperatures that could eliminate 56,000 plant and 3,700 animal species.

Mass Extinctions

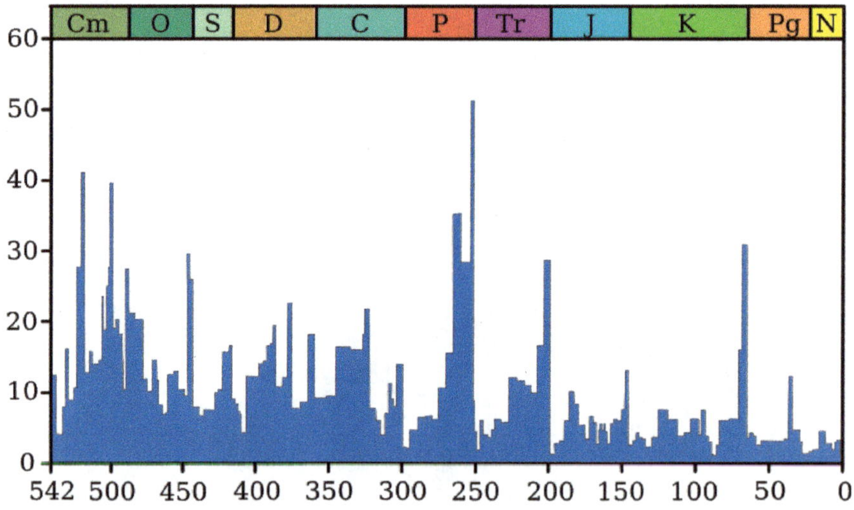

The blue graph shows the apparent *percentage* (not the absolute number) of marine animal genera becoming extinct during any given time interval. It does not represent all marine species, just those that are readily fossilized. The labels of the "Big Five" extinction events are clickable hyperlinks; see Extinction event for more details. *(source and image info)*

There have been at least five mass extinctions in the history of life on earth, and four in the last 350 million years in which many species have disappeared in a relatively short period of geological time. A massive eruptive event is considered to be one likely cause of the "Permian–Triassic extinction event" about 250 million years ago. This was estimated to have killed 90% of species then existing. There is also evidence to suggest that this event was preceded by another mass extinction, known as Olson's Extinction. The Cretaceous–Paleogene extinction event (K-Pg) occurred 66 million years ago, at the end of the Cretaceous period. It is best known for having wiped out non-avian dinosaurs, among many other species.

Modern Extinctions

According to a 1998 survey of 400 biologists conducted by New York's American Museum of Natural History, nearly 70% believed that the Earth is currently in the early stages of a human-caused mass extinction, known as the Holocene extinction. In that survey, the same proportion of respondents agreed with the prediction that up to 20% of all living populations could become extinct within 30 years (by 2028). Biologist E. O. Wilson estimated in 2002 that if current rates of human destruction of the biosphere continue, one-half of all plant and animal species of life on earth will be extinct in 100 years. More significantly, the current rate of global species extinctions is estimated as 100 to 1000 times background rates (the average extinction rates in the evolutionary time scale of planet Earth), while future rates are likely 10,000 times higher. However, some groups are going extinct much faster.

History of Scientific Understanding

Dilophosaurus, one of the many extinct dinosaur genera. The cause of the Cretaceous–Paleogene extinction event is a subject of much debate amongst researchers

When it was first described in the 1750s, the idea of extinction was threatening to those who held a belief in the great chain of being, a theological position that did not allow for "missing links".

The possibility of extinction was not widely accepted before the 1800s. The devoted naturalist, Carl Linnaeus, could "hardly entertain" the idea that humans could cause the extinction of a species. When parts of the world had not been thoroughly examined and charted, scientists could not rule out that animals found only in the fossil record were not simply hiding in unexplored regions of the Earth. Georges Cuvier is credited with establishing extinction as a fact in a 1796 lecture to the French Institute. Cuvier's observations of fossil bones convinced him that they did not originate in extant animals. This discovery was critical for the spread of uniformitarianism, and led to the first book publicizing the idea of evolution though Cuvier himself strongly opposed the theories of evolution advanced by Lamarck and others.

Human Attitudes and Interests

Extinction is an important research topic in the field of zoology, and biology in general, and has also become an area of concern outside the scientific community. A number of organizations, such as the Worldwide Fund for Nature, have been created with the goal of preserving species from extinction. Governments have attempted, through enacting laws, to avoid habitat destruction, agricultural over-harvesting, and pollution. While many human-caused extinctions have been accidental, humans have also engaged in the deliberate destruction of some species, such as dangerous viruses, and the total destruction of other problematic species has been suggested. Other species were deliberately driven to extinction, or nearly so, due to poaching or because they were undesirable, or to push for other human agendas. One example was the near extinction of the American bison, which was nearly wiped out by mass hunts sanctioned by the United States government, to force the removal of Native Americans, many of whom relied on the bison for food.

Biologist Bruce Walsh of the University of Arizona states three reasons for scientific interest in the preservation of species; genetic resources, ecosystem stability, and ethics; and today the scientific community stress[es] the importance of maintaining biodiversity.

In modern times, commercial and industrial interests often have to contend with the effects of production on plant and animal life. However, some technologies with minimal, or no, proven harmful effects on *Homo sapiens* can be devastating to wildlife (for example, DDT). BiogeographerJared Diamond notes that while big business may label environmental concerns as "exaggerated", and often cause "devastating damage", some corporations find it in their interest to adopt good conservation practices, and even engage in preservation efforts that surpass those taken by national parks.

Governments sometimes see the loss of native species as a loss to ecotourism, and can enact laws with severe punishment against the trade in native species in an effort to prevent extinction in the wild. Nature preserves are created by governments as a means to provide continuing habitats to species crowded by human expansion. The 1992 Convention on Biological Diversity has resulted in international Biodiversity Action Plan programmes, which attempt to provide comprehensive guidelines for government biodiversity conservation. Advocacy groups, such as The Wildlands Project and the Alliance for Zero Extinctions, work to educate the public and pressure governments into action.

People who live close to nature can be dependent on the survival of all the species in their environment, leaving them highly exposed to extinction risks. However, people prioritize day-to-day survival over species conservation; with human overpopulation in tropical developing countries, there has been enormous pressure on forests due to subsistence agriculture, including slash-and-burn agricultural techniques that can reduce endangered species's habitats.

Planned Extinction

Completed

- The smallpox virus is now extinct in the wild, although samples are retained in laboratory settings.
- The rinderpest virus, which infected domestic cattle, is now extinct in the wild.

Proposed

The poliovirus is now confined to small parts of the world due to extermination efforts.

Dracunculus medinensis, a parasitic worm which causes the disease dracunculiasis, is now close to eradication thanks to efforts led by the Carter Center.

Treponema pallidum pertenue, a bacterium which causes the disease yaws, is in the process of being eradicated.

Biologist Olivia Judson has advocated the deliberate extinction of certain disease-carrying mosquito species. In a September 25, 2003 *New York Times* article, she advocated "specicide of thirty mosquito species by introducing a genetic element which can insert itself into another crucial gene, to create recessive "knockout genes. She says that the *Anopheles* mosquitoes (which spread malaria) and *Aedes* mosquitoes (which spread dengue fever, yellow fever, elephantiasis, and other diseases) represent only 30 species; eradicating these would save at least one million human lives per annum, at a cost of reducing the genetic diversity of the family Culicidae by only 1%. She further argues that since species become extinct "all the time" the disappearance of a few more will not destroy the ecosystem: We're not left with a wasteland every time a species vanishes. Removing one species sometimes causes shifts in the populations of other species—but different need not mean worse. In addition, anti-malarial and mosquito control programs offer little realistic hope to the 300 million people in developing nations who will be infected with acute illnesses this year. Although trials are ongoing, she writes that if they fail: "We should consider the ultimate swatting."

Cloning

Some, such as Harvard geneticist George M. Church, believe that ongoing technological advances will let us "bring back to life" an extinct species by cloning, using DNA from the remains of that species. Proposed targets for cloning include the mammoth, the thylacine, and the Pyrenean ibex. For this to succeed, enough individuals would have to be cloned, from the DNA of different individuals (in the case of sexually reproducing organisms) to create a viable population. Though bioethical and philosophical objections have been raised, the cloning of extinct creatures seems theoretically possible.

In 2003, scientists tried to clone the extinct Pyrenean ibex (*C. p. pyrenaica*). This attempt failed: of the 285 embryos reconstructed, 54 were transferred to 12 mountain goats and mountain goat-domestic goat hybrids, but only two survived the initial two months of gestation before they too died. In 2009, a second attempt was made to clone the Pyrenean ibex: one clone was born alive, but died seven minutes later, due to physical defects in the lungs.

SPECIATION

Speciation is the evolutionary process by which reproductively isolated biological populations evolve to become distinct species. The biologist Orator F. Cook coined the term 'speciation' in 1906 for the splitting of lineages or "cladogenesis," as opposed to "anagenesis" or "phyletic evolution" within lineages. Charles Darwin was the first to describe the role of natural selection in speciation in his 1859 book The Origin of Species. He also identified sexual selection as a likely mechanism, but found it problematic.

There are four geographic modes of speciation in nature, based on the extent to which speciating populations are isolated from one another: allopatric, peripatric, parapatric, and sympatric. Speciation may also be induced artificially, through animal husbandry, agriculture, or laboratory experiments. Whether genetic drift is a minor or major contributor to speciation is the subject matter of much ongoing discussion.

Modes of Speciation

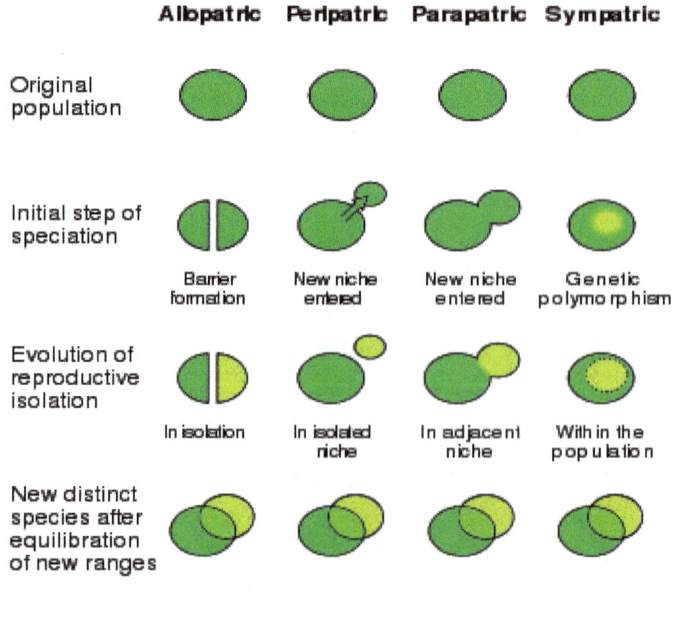

Comparison of allopatric, peripatric, parapatric and sympatric speciation.

All forms of natural speciation have taken place over the course of evolution; however, debate persists as to the relative importance of each mechanism in driving biodiversity.

One example of natural speciation is the diversity of the three-spined stickleback, a marine fish that, after the last glacial period, has undergone speciation into new freshwater colonies in isolated lakes and streams. Over an estimated 10,000 generations, the sticklebacks show structural differences that are greater than those seen between different genera of fish including variations in fins, changes in the number or size of their bony plates, variable jaw structure, and color differences.

Allopatric

During allopatric (from the ancient Greek *allos*, "other" + Greek *patr* , "fatherland") speciation, a population splits into two geographically isolated populations (for example, by habitat fragmentation due to geographical change such as mountain formation). The isolated populations then undergo genotypic and/or phenotypic divergence as: (a) they become subjected to dissimilar selective pressures; (b) they independently undergo genetic drift; (c) different mutations arise in the two populations. When the populations come back into contact, they have evolved such that they are reproductively isolated and are no longer capable of exchanging genes. Island genetics is the term associated with the tendency of small, isolated genetic pools to produce unusual traits. Examples include insular dwarfism and the radical changes among certain famous island chains, for example on Komodo. The Galápagos Islands are particularly famous for their influence on Charles Darwin. During his five weeks there he heard that Galápagos tortoises could be identified by island, and noticed that finches differed from one island to another, but it was only nine months later that he reflected that such facts could show that species were changeable. When he returned to England, his speculation on evolution deepened after experts informed him that these were separate species, not just varieties, and famously that other differing Galápagos birds were all species of finches. Though the finches were less important for Darwin, more recent research has shown the birds now known as Darwin's finches to be a classic case of adaptive evolutionary radiation.

Peripatric

In peripatric speciation, a subform of allopatric speciation, new species are formed in isolated, smaller peripheral populations that are prevented from exchanging genes with the main population. It is related to the concept of a founder effect, since small populations often undergo bottlenecks. Genetic drift is often proposed to play a significant role in peripatric speciation.

Case Studies:

- Mayr bird fauna
- The Australian bird *Petroica multicolor*
- Reproductive isolation occurs in populations of *Drosophila* subject to population bottlenecking

Parapatric

In parapatric speciation, there is only partial separation of the zones of two diverging populations afforded by geography; individuals of each species may come in contact or cross habitats from time to time, but reduced fitness of the heterozygote leads to selection for behaviours or mechanisms that prevent their interbreeding. Parapatric speciation is modelled on continuous variation within a single, connected habitat acting as a source of natural selection rather than the effects of isolation of habitats produced in peripatric and allopatric speciation.

Parapatric speciation may be associated with differential landscape-dependent selection. Even if there is a gene flow between two populations, strong differential selection may impede assimilation and different species may eventually develop. Habitat differences may be more important in the development of reproductive isolation than the isolation time. Caucasian rock lizards *Darevskia rudis*, *D. valentini* and *D. portschinskii* all hybridize with each other in their hybrid zone; however, hybridization is stronger between *D. portschinskii* and *D. rudis*, which separated earlier but live in similar habitats than between *D. valentini* and two other species, which separated later but live in climatically different habitats.

Ecologists refer to parapatric and peripatric speciation in terms of ecological niches. A niche must be available in order for a new species to be successful.

Case Studies:

- Ring species
    - The *Larus* gulls form a ring species around the North Pole.
    - The *Ensatina* salamanders, which form a ring round the Central Valley in California.
    - The greenish warbler (*Phylloscopus trochiloides*), around the Himalayas.
- The grass *Anthoxanthum* has been known to undergo parapatric speciation in such cases as mine contamination of an area.

Sympatric

Japanese rice fish

Freshwater angelfish, a cichlid

Sympatric speciation refers to the formation of two or more descendant species from a single ancestral species all occupying the same geographic location.

Often-cited examples of sympatric speciation are found in insects that become dependent on different host plants in the same area. However, the existence of sympatric speciation as a mechanism of speciation remains highly debated.

The best illustrated example of sympatric speciation is that of the cichlids of East Africa inhabiting the Rift Valley lakes, particularly Lake Victoria, Lake Malawi and Lake Tanganyika. There are over 800 described species, and according to estimate, there could be well over 1,600 species in the region. All the species have diversified from a common ancestral fish, the Japanese rice fish (*Oryzias latipes*) about 113 million years ago. Their evolution is cited as an example of both natural and sexual selection. A 2008 study suggests that sympatric speciation has occurred in Tennessee cave salamanders. Sympatric speciation driven by ecological factors may also account for the extraordinary diversity of crustaceans living in the depths of Siberia's Lake Baikal.

Budding speciation has been proposed as a particular form of sympatric speciation, whereby small groups of individuals become progressively more isolated from the ancestral stock by breeding preferentially with one another. This type of speciation would be driven by the conjunction of various advantages of inbreeding such as the expression of advantageous recessive phenotypes, reducing the recombination load, and reducing the cost of sex

The hawthorn fly (*Rhagoletis pomonella*), also known as the apple maggot fly, appears to be undergoing sympatric speciation. Different populations of hawthorn fly feed on different fruits. A distinct population emerged in North America in the 19th century some time after apples, a non-native species, were introduced. This apple-feeding population normally feeds only on apples and not on the historically preferred fruit of hawthorns. The current hawthorn feeding population does not normally feed on apples. Some evidence, such as that six out of thirteen allozyme loci are different, that hawthorn flies mature later in the season and take longer to mature than apple flies; and that there is little evidence of interbreeding (researchers have documented a 4-6% hybridization rate) suggests that sympatric speciation is occurring.

Rhagoletis pomonella

Reinforcement

Reinforcement, also called the *Wallace effect*, is the process by which natural selection increases reproductive isolation. It may occur after two populations of the same species are separated and then come back into contact. If their reproductive isolation was complete, then they will have already developed into two separate incompatible species. If their reproductive isolation is incomplete, then further mating between the populations will produce hybrids, which may or may not be fertile. If the hybrids are infertile, or fertile but less fit than their ancestors, then there will be further reproductive isolation and speciation has essentially occurred (e.g., as in horses and donkeys.)

The reasoning behind this is that if the parents of the hybrid offspring each have naturally selected traits for their own certain environments, the hybrid offspring will bear traits from both, therefore would not fit either ecological niche as well as either parent. The low fitness of the hybrids would cause selection to favor assortative mating, which would control hybridization. This is sometimes called the Wallace effect after the evolutionary biologist Alfred Russel Wallace who suggested in the late 19th century that it might be an important factor in speciation. Conversely, if the hybrid offspring are more fit than their ancestors, then the populations will merge back into the same species within the area they are in contact.

Reinforcement favoring reproductive isolation is required for both parapatric and sympatric speciation. Without reinforcement, the geographic area of contact between different forms of the same species, called their hybrid zone, will not develop into a boundary between the different species. Hybrid zones are regions where diverged populations meet and interbreed. Hybrid offspring are very common in these regions, which are usually created by diverged species coming into secondary contact. Without reinforcement, the two species would have uncontrollable inbreeding. Reinforcement may be induced in artificial selection experiments as described below.

Ecological and Parallel Speciation

Ecological selection is “the interaction of individuals with their environment during resource acquisition”. Natural selection is inherently involved in the process of speciation, whereby, “under ecological speciation, populations in different environments, or populations exploiting different resources, experience contrasting natural selection pressures on the traits that directly or indirectly bring about the evolution of reproductive isolation”. Evidence for the role ecology plays in the process of speciation exists. Studies of stickleback populations support ecologically-linked speciation arising as a by-product, alongside numerous studies of parallel speciation.

Parallel speciation is where “greater reproductive isolation repeatedly evolves between independent populations adapting to contrasting environments than between independent populations adapting to similar environments”. It is established that ecological speciation occurs and with much of the evidence, “...accumulated from top-down studies of adaptation and reproductive isolation”.

Sexual Selection

It is widely appreciated that sexual selection could drive speciation in many clades, independently of natural selection. However the term “speciation”, in this context, tends to be used in two different, but not mutually exclusive senses. The first and most commonly used sense refers to the “birth” of new species. That is, the splitting of an existing species into two separate species, or the budding off of a new species from a parent species, both driven by a biological fashion fad (a preference for a feature, or features, in one or both sexes, that do not necessarily have any adaptive qualities). In the second sense, speciation refers the wide-spread tendency of sexual creatures to be grouped into clearly defined species, rather than forming a continuum of phenotypes both in time and space - which would be the more obvious or logical consequence of natural selection. This was indeed recognized by Darwin as problematic, and included in his *On the Origin of Species* (1859), under the heading “Difficulties with the Theory”. There are several suggestions as to how mate choice might play a significant role in resolving Darwin’s dilemma.

Artificial Speciation

European mouflon (*Ovis aries musimon*)

Male *Drosophila pseudoobscura*

New species have been created by domesticated animal husbandry, but the initial dates and methods of the initiation of such species are not clear. Often, the domestic counterpart of the wild ancestor can still interbreed and produce fertile offspring as in the case of domestic cattle, that can be considered the same species as several varieties of wild ox, gaur, yak, etc., or domestic sheep that can interbreed with the mouflon.

The best-documented creations of new species in the laboratory were performed in the late 1980s. William R. Rice and George W. Salt bred *Drosophila melanogaster* fruit flies using a maze with three different choices of habitat such as light/dark and wet/dry. Each generation was placed into the maze, and the groups of flies that came out of two of the eight exits were set apart to breed with each other in their respective groups. After thirty-five generations, the two groups and their offspring were isolated reproductively because of their strong habitat preferences: they mated only within the areas they preferred, and so did not mate with flies that preferred the other areas. The history of such attempts is described by Rice and Elen E. Hostert (1993).

Diane Dodd used a laboratory experiment to show how reproductive isolation can evolve in *Drosophila pseudoobscura* fruit flies after several generations by placing them in different media, starch- and maltose-based media.

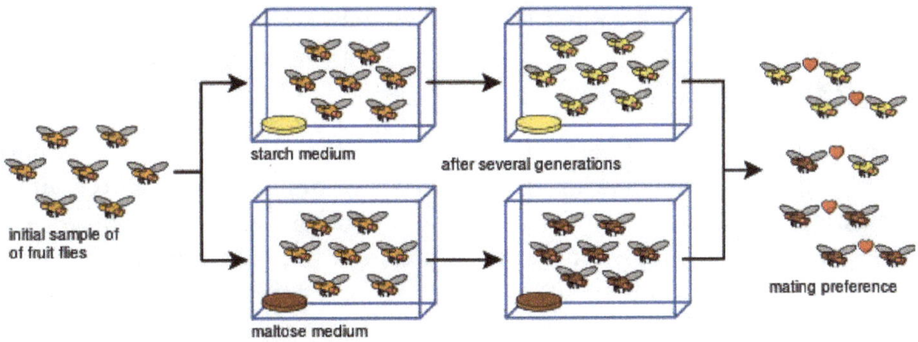

Dodd's experiment has been easy for many others to replicate, including with other kinds of fruit flies and foods. Research in 2005 has shown that this rapid evolution of reproductive isolation may in fact be a relic of infection by *Wolbachia* bacteria.

Alternatively, these observations are consistent with the notion that sexual creatures are inherently reluctant to mate with individuals whose appearance or behavior is different from the norm. The risk that such deviations are due to heritable maladaptations is very high. Thus, if a sexual creature, unable to predict natural selection's future direction, is conditioned to produce the fittest offspring possible, it will avoid mates with unusual habits or features. Sexual creatures will then inevitably tend to group themselves into reproductively isolated species.

Genetics

Few speciation genes have been found. They usually involve the reinforcement process of late stages of speciation. In 2008, a speciation gene causing reproductive isolation was reported. It causes hybrid sterility between related subspecies. The order of speciation of three groups from a common ancestor may be unclear or unknown; a collection of three such species is referred to as a "trichotomy."

Speciation Via Polyploidization

Polyploidy is a mechanism that has caused many rapid speciation events in sympatry because offspring of, for example, tetraploid x diploid matings often result in triploid sterile progeny. However, not all polyploids are reproductively isolated from their parental plants, and gene flow may still occur for example through triploid hybrid x diploid matings that produce tetraploids, or matings between meiotically unreduced gametes from diploids and gametes from tetraploids (see also hybrid speciation).

It has been suggested that many of the existing plant and most animal species have undergone an event of polyploidization in their evolutionary history. Reproduction of successful polyploid species is sometimes asexual, by parthenogenesis or apomixis, as for unknown reasons many asexual organisms are polyploid. Rare instances of polyploid mammals are known, but most often result in prenatal death.

Hybrid Speciation

Hybridization between two different species sometimes leads to a distinct phenotype. This phenotype can also be fitter than the parental lineage and as such natural selection may then favor these individuals. Eventually, if reproductive isolation is achieved, it may lead to a separate species. However, reproductive isolation between hybrids and their parents is particularly difficult to achieve and thus hybrid speciation is considered an extremely rare event. The mariana mallard is thought to have arisen from hybrid speciation.

Hybridization is an important means of speciation in plants, since polyploidy (having more than two copies of each chromosome) is tolerated in plants more readily than in animals. Polyploidy is important in hybrids as it allows reproduction, with the two different sets of chromosomes each being able to pair with an identical partner during meiosis. Polyploids also have more genetic diversity, which allows them to avoid inbreeding depression in small populations.

Hybridization without change in chromosome number is called homoploid hybrid speciation. It is considered very rare but has been shown in *Heliconius* butterflies and sunflowers. Polyploid speciation, which involves changes in chromosome number, is a more common phenomenon, especially in plant species.

Gene Transposition

Theodosius Dobzhansky, who studied fruit flies in the early days of genetic research in 1930s, speculated that parts of chromosomes that switch from one location to another might cause a species to split into two different species. He mapped out how it might be possible for sections of chromosomes to relocate themselves in a genome. Those mobile sections can cause sterility in inter-species hybrids, which can act as a speciation pressure. In theory, his idea was sound, but scientists long debated whether it actually happened in nature. Eventually a competing theory involving the gradual accumulation of mutations was shown to occur in nature so often that geneticists largely dismissed the moving gene hypothesis.

However, 2006 research shows that jumping of a gene from one chromosome to another can contribute to the birth of new species. This validates the reproductive isolation mechanism, a key component of speciation.

Historical Background

In addressing the question of the origin of species, there are two key issues: (1) what are the evolutionary mechanisms of speciation, and (2) what accounts for the separateness and individuality of species in the biota? Since Charles Darwin's time, efforts to understand the nature of species have primarily focused on the first aspect, and it is now widely agreed that the critical factor behind the origin of new species is reproductive isolation. Next we focus on the second aspect of the origin of species.

Darwin's Dilemma: why do Species Exist?

In *On the Origin of Species* (1859), Darwin interpreted biological evolution in terms of natural selection, but was perplexed by the clustering of organisms into species. Chapter 6 of Darwin's book is entitled "Difficulties of the Theory." In discussing these "difficulties" he noted "Firstly, why, if species have descended from other species by insensibly fine gradations, do we not everywhere see innumerable transitional forms? Why is not all nature in confusion instead of the species being, as we see them, well defined?" This dilemma can be referred to as the absence or rarity of transitional varieties in habitat space.

Another dilemma, related to the first one, is the absence or rarity of transitional varieties in time. Darwin pointed out that by the theory of natural selection innumerable transitional forms must have existed, and wondered why do we not find them embedded in countless numbers in the crust of the earth. That clearly defined species actually do exist in nature in both space and time implies that some fundamental feature of natural selection operates to generate and maintain species.

The Effect of Sexual Reproduction on Species Formation

It has been argued that the resolution of Darwin's first dilemma lies in the fact that out-crossing sexual reproduction has an intrinsic cost of rarity. The cost of rarity arises as follows. If, on a resource gradient, a large number of separate species evolve, each exquisitely adapted to a very narrow band on that gradient, each species will, of necessity, consist of very few members. Finding a mate under these circumstances may present difficulties when many of the individuals in the neighborhood belong to other species. Under these circumstances, any species that happens, by chance, to increase in population size (at the expense of one or other of its neighboring species, if the environment is saturated), this will immediately make it easier for its members to find sexual partners. The members of the neighboring species, whose population sizes have decreased, experience greater difficulty in finding mates, and therefore form pairs less frequently than in the larger species. This has a snowball effect, with large species growing at the expense of the smaller, rarer species, eventually driving them to extinction. Eventually, only a few species remain, each distinctly different from the other. The cost of rarity not only involves the costs of failure to find a mate, but also indirect costs such as the cost of communication in seeking out a partner at low population densities.

African pygmy kingfisher, showing coloration shared by all adults of that species to a high degree of fidelity.

Rarity brings with it other costs. Rare and unusual features are very seldom advantageous. In most instances, they indicate a (non-silent mutation), almost certain to be deleterious. It therefore behooves sexual creatures to avoid mates sporting rare or unusual features. Sexual populations therefore rapidly shed rare or peripheral phenotypic features, thus canalizing the entire external appearance, as illustrated in the accompanying illustration of the African pygmy kingfisher, *Ispidina picta*. This remarkable uniformity of all the adult members of a sexual species has stimulated the proliferation of field guides on birds, mammals, reptiles, insects, and many other taxa, in which a species can be described with a single illustration (or two, in the case of sexual dimorphism). Once a population has become as homogeneous in appearance as is typical of most species (and is illustrated in the photograph of the African pygmy kingfisher), its members will avoid mating with members of other populations that look different from themselves. Thus, the avoidance of mates displaying rare and unusual phenotypic features inevitably leads to reproductive isolation, one of the hallmarks of speciation.

In the contrasting case of organisms that reproduce asexually, there is no cost of rarity; consequently, there are only benefits to fine-scale adaptation. Thus, asexual organisms very frequently show the continuous variation in form (often in many different directions) that Darwin expected evolution to produce, making their classification into species (more correctly, morphospecies) very difficult.

Rates of Speciation

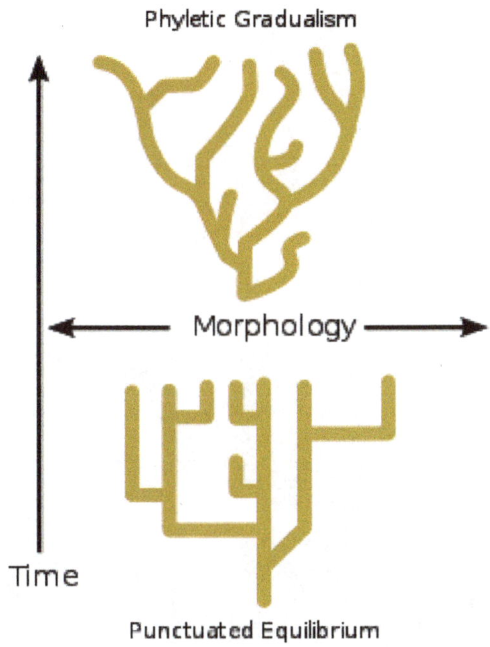

Phyletic gradualism, above, consists of relatively slow change over geological time. Punctuated equilibrium, bottom, consists of morphological stability and rare, relatively rapid bursts of evolutionary change (see the illustration at the bottom of the page).

There is debate as to the rate at which speciation events occur over geologic time. While some evolutionary biologists claim that speciation events have remained relatively constant and gradual over time (known as Phyletic gradualism - see diagram), some palaeontologists such as Niles Eldredge and Stephen Jay Gould have argued that species usually remain unchanged over long stretches of time, and that speciation occurs only over relatively brief intervals, a view known as *punctuated equilibrium*. (See diagram, and Darwin's dilemma.)

Punctuated Evolution

Evolution can be extremely rapid, as shown in the creation of domesticated animals and plants in a very short geological space of time, spanning only a few tens of thousands of years. Maize (*Zea mays*), for instance, was created in Mexico in only a few thousand years, starting about 7,000 to 12,000 years ago. This raises the question of why the long term rate of evolution is far slower than is theoretically possible.

Plants **and** domestic animals **can differ markedly from their wild ancestors**

Top: wild teosinte; middle: maize-teosinte hybrid; bottom: maize

Ancestral wild cabbage

Domesticated cauliflower

Ancestral Prussian carp

Domestic goldfish

Ancestral mouflon

Domestic sheep

Evolution is imposed on species or groups. It is not planned or striven for in some Lamarckist way. The mutations on which the process depends are random events, and, except for the "silent mutations which do not affect the functionality or appearance of the carrier, are thus usually disadvantageous, and their chance of proving to be useful in the future is vanishingly small. Therefore, while a species or group might benefit from being able to adapt to a new environment by accumulating a wide range of genetic variation, this is to the detriment of the *individuals* who have to carry these mutations until a small, unpredictable minority of them ultimately contributes to such an adaptation. Thus, the *capability* to evolve would require group selection, a concept discredited by (for example) George C. Williams, John Maynard Smith and Richard Dawkins as selectively disadvantageous to the individual.

The resolution to Darwin's second dilemma might thus come about as follows:

If sexual individuals are disadvantaged by passing mutations on to their offspring, they will avoid mutant mates with strange or unusual characteristics. Mutations that affect the external appearance of their carriers will then rarely be passed on to the next and subsequent generations. They would therefore seldom be tested by natural selection. Evolution is, therefore, effectively halted or slowed down considerably. The only mutations that can accumulate in a population, on this punctuated equilibrium view, are ones that have no noticeable effect on the outward appearance and functionality of their bearers (i.e., they are silent or neutral mutations, which can be, and are, used to trace the relatedness and age of populations and species.) This argument implies that evolution can only occur if mutant mates cannot be avoided, as a result of a severe scarcity of potential mates. This is most likely to occur in small, isolated communities. These occur most commonly on small islands, in remote valleys, lakes, river systems, or caves, or during the aftermath of a mass extinction. Under these circumstances, not only is the choice of mates severely restricted but population bottlenecks, founder effects, genetic drift and inbreeding cause rapid, random changes in the isolated population's genetic composition. Furthermore, hybridization with a related species trapped in the same isolate might introduce additional genetic changes. If an isolated population such as this survives its genetic upheavals, and subsequently expands into an unoccupied niche, or into a niche in which it has an advantage over its competitors, a new species, or subspecies, will have come in being. In geological terms this will be an abrupt event. A resumption of avoiding mutant mates will thereafter result, once again, in evolutionary stagnation.

In apparent confirmation of this punctuated equilibrium view of evolution, the fossil record of an evolutionary progression typically consists of species that suddenly appear, and ultimately disappear, hundreds of thousands or millions of years later, without any change in external appearance. Graphically, these fossil species are represented by horizontal lines, whose lengths depict how long each of them existed. The horizontality of the lines illustrates the unchanging appearance of each of the fossil species depicted on the graph. During each species' existence new species appear at random intervals, each also lasting many hundreds of thousands of years before disappearing without a change in appearance. The exact relatedness of these concurrent species is generally impossible to determine. This is illustrated in the following diagram depicting the evolution of modern humans from the time that the hominins separated from the line that led to the evolution of our closest living primate relatives, the chimpanzees.

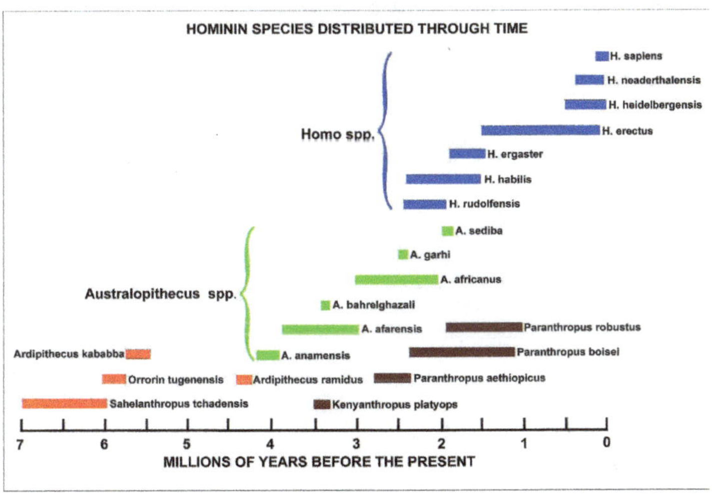

Distribution of Hominin species over time

For similar evolutionary time lines see, for instance, the paleontological list of African dinosaurs, Asian dinosaurs, the Lampriformes and Amiiformes.

BACKGROUND EXTINCTION RATE

Background extinction rate, also known as the normal extinction rate, refers to the standard rate of extinction in earth's geological and biological history before humans became a primary contributor to extinctions. This is primarily the pre-human extinction rates during periods in between major extinction events.

Extinctions are a normal part of the evolutionary process, and the background extinction rate is a measurement of "how often" they naturally occur. Normal extinction rates are often used as a comparison to present day extinction rates, to illustrate the higher frequency of extinction today than in all periods of non-extinction events before it.

Background extinction rates have not remained constant, although changes are measured over geological time, covering millions of years.

Measurement

Background extinction rates are typically measured three different ways. The first is simply the number of species that normally go extinct over a given period of time. For example, at the background rate one species of bird will go extinct every estimated 400 years. Another way the extinction rate can be given is in million species years (MSY). For example, there is approximately one extinction estimated per million species years. From a purely mathematical standpoint this means that if there are a million species on the planet earth, one would go extinct every year, while if there was only one species it would go extinct in one million years, etc. The third way is in giving species survival rates over time. For example, given normal extinction rates species typically exist for 5–10 hundred thousand years before going extinct.

Lifespan Estimates

Some species lifespan estimates by taxonomy

Taxonomy	Source of Estimate	Species Average Lifespan (Millions of Years)
All Invertebrates	Raup (1978)	11
Marine Invertebrates	Valentine (1970)	5–10
Marine Animals	Raup (1991)	4
Marine Animals	Sepkoski (1992)	5
All Fossil Groups	Simpson (1952)	0.5–5
Mammals	Martin (1993)	1
Cenozoic Mammals	Raup and Stanley (1978)	1–2
Diatoms	Van Valen	8
Dinoflagelates	Van Valen (1973)	13
Planktonic Foraminifera	Van Valen (1973)	7
Cenozoic Bivalves	Raup and Stanley (1978)	10
Echinoderms	Durham (1970)	6
Silurian Graptolites	Rickards (1977)	2

Adapted from the book Extinction Rates, edited by J. Lawton, and R. May

Accuracy

The fact that we do not currently know the total number of species, in the past nor the present, makes it very difficult to accurately calculate the non-anthropogenically influenced extinction rates. As a rate, it is essential to know not just the number of extinctions, but also the number of non-extinctions. This fact, coupled with the fact that the rates do not remain constant, significantly reduces accuracy in estimates of the normal rate of extinctions.

Drivers That Affect Biodiversity

Habitat change (see: habitat fragmentation or habitat destruction) is the most important driver currently affecting biodiversity, as some 40% of forests and ice-free habitats have been converted to cropland or pasture. Other drivers are: overexploitation, pollution, invasive species, and climate change.

Measuring Diversity

Biodiversity is usually plotted as the richness of a geographic area, with some reference to a temporal scale. Types of biodiversity include taxonomic or species, ecological, morphological, and genetic diversity. Taxonomic diversity, that is the number of species, genera, family is the most commonly assessed type. A few studies have attempted to quantitatively clarify the relationship between different types of diversity. For example, the biologist Sarda Sahney has found a close link between vertebrate taxonomic and ecological diversity.

Known Species

Insects make up the vast majority of animal species.

Chapman, 2005 and 2009 has attempted to compile perhaps the most comprehensive recent statistics on numbers of extant species, drawing on a range of published and unpublished sources, and has come up with a figure of approximately 1.9 million estimated described taxa, as against possibly a total of between 11 and 12 million anticipated species overall (described plus undescribed), though other reported values for the latter vary widely. It is important to note that in many cases, the values given for "Described" species are an estimate only (sometimes a mean of reported figures in the literature) since for many of the larger groups in particular, comprehensive lists of valid species names do not currently exist. For fossil species, exact or even approximate numbers are harder to find; Raup, 1986 includes data based on a compilation of 250,000 fossil species so the true number is undoubtedly somewhat higher than this. It should also be noted that the number of described species is increasing by around 18,000-19,000 extant, and approaching 2,000 fossil species each year at the present time, The number of published species *names* is higher than the number of described species, sometimes considerably so, on account of the publication, through time, of multiple names (synonyms) for the same accepted taxon in many cases.

Based on Chapman's (2009) report, the estimated numbers of described extant species as of 2009 can be broken down as follows:

Major group	Described	Global estimate (described + undescribed)	Component group	Described	Global estimate	Component group (2)	Described	Global estimate
Chordates	64,788	~80,500						
			Mammals	5,487	~5,500			
			Birds	9,990	>10,000			
			Reptiles	8,734	~10,000			
			Amphibia	6,515	~15,000			
			Fishes	31,153	~40,000			
			Agnatha	116	unknown			
			Cephalochordata	33	unknown			
			Tunicata	2,760	unknown			
Invertebrates	~1,359,365	~6,755,830						
			Hemichordata	108	~110			
			Echinodermata	7,003	~14,000			
			Insecta	~1,000,000 (965,431-1,015,897)	~5,000,000			
						Archaeognatha	470	
						Blattodea	3,684-4,000	
						Coleoptera	360,000-~400,000	1,100,000
						Dermaptera	1,816	
						Diptera	152,956	240,000
						Embioptera	200-300	2,000

						Ephemeroptera	2,500-<3,000	
						Grylloblattaria	24	
						Hemiptera	80,000-88,000	
						Hymenoptera	115,000	>300,000
						Isoptera	2,600-2,800	4,000
						Lepidoptera	174,250	300,000-500,000
						Mantodea	2,200	
						Mecoptera	481	
						Megaloptera	250-300	
						Neuroptera	~5,000	
						Odonata	6,500	
						Orthoptera	24,380	
						Phasmatodea (Phasmida)	2,500-3,300	
						Phthiraptera	>3,000-~3,200	
						Plecoptera	2,274	
						Psocoptera	3,200-~3,500	
						Siphonaptera	2,525	
						Strepsiptera	596	
						Thysanoptera	~6,000	
						Trichoptera	12,627	
						Zoraptera	28	
						Zygentoma (Thysanura)	370	
			Arachnida	102,248	~600,000			
			Pycnogonida	1,340	unknown			
			Myriapoda	16,072	~90,000			
			Crustacea	47,000	150,000			
			Onychophora	165	~220			
			non-Insect Hexapoda	9,048	52,000			
			Mollusca	~85,000	~200,000			
			Annelida	16,763	~30,000			
			Nematoda	<25,000	~500,000			
			Acanthocephala	1,150	~1,500			
			Platyhelminthes	20,000	~80,000			
			Cnidaria	9,795	unknown			

			Porifera	~6,000	~18,000			
			Other Invertebrates	12,673	~20,000			
						Placozoa	1	-
						Monoblastozoa	1	-
						Mesozoa (Rhombozoa, Orthonectida)	106	-
						Ctenophora	166	200
						Nemertea (Nemertina)	1,200	5,000-10,000
						Rotifera	2,180	-
						Gastrotricha	400	-
						Kinorhyncha	130	-
						Nematomorpha	331	~2,000
						Entoprocta (Kamptozoa)	170	170
						Gnathostomulida	97	-
						Priapulida	16	-
						Loricifera	28	>100
						Cycliophora	1	-
						Sipuncula	144	-
						Echiura	176	-
						Tardigrada	1,045	-
						Phoronida	10	-
						Ectoprocta (Bryozoa)	5,700	~5,000
						Brachiopoda	550	-
						Pentastomida	100	-
						Chaetognatha	121	-
Plants sens. lat.	~310,129	~390,800						
			Bryophyta	16,236	~22,750			
						Liverworts	~5,000	~7,500
						Hornworts	236	~250
						Mosses	~11,000	~15,000
			Algae (Plant)	12,272	unknown			
						Charophyta	2,125	-
						Chlorophyta	4,045	-
						Glaucophyta	5	-
						Rhodophyta	6,097	-

			Vascular Plants	281,621	~368,050			
						Ferns and allies	~12,000	~15,000
						Gymnosperms	~1,021	~1,050
						Magnoliophyta	~268,600	~352,000
Fungi	98,998 (incl. Lichens 17,000)	1,500,000 (incl. Lichens ~25,000)						
Others	~66,307	~2,600,500						
			Chromista [incl. brown algae, diatoms and other groups]	25,044	~200,500			
			Protoctista [i.e. residual protist groups]	~28,871	>1,000,000			
			Prokaryota [Bacteria and Archaea, excl. Cyanophyta]	7,643	~1,000,000			
			Cyanophyta	2,664	unknown			
			Viruses	2,085	400,000			
Total (2009 data)	1,899,587	~11,327,630						

Estimates of Total Number of Species

However the *total* number of species for some taxa may be much higher.

- 10-30 million insects;
- 5-10 million bacteria;
- 1.5 million fungi;
- ~1 million mites
- ~1 million protists

In 1982, Terry Erwin published an estimate of global species richness of 30 million, by extrapolating from the numbers of beetles found in a species of tropical tree. In one species of tree, Erwin identified 1200 beetle species, of which he estimated 163 were found only in that type of tree. Given the 50,000 described tropical tree species, Erwin suggested that there are almost 10 million beetle species in the tropics. In 2011 a study published in PLoS Biology estimated there to be 8.7 million ± 1.3 million eukaryotic species on Earth. A 2016 study concludes that Earth is home to 1 trillion species.

Global Biodiversity Indices

After the Convention on Biological Diversity was signed in 1992, biological conservation became a priority for the international community. There are several indicators used that describe trends in global biodiversity. However, there is no single indicator for all extant species as not all have been described and measured over time. There are different ways to measure changes in biodiversity. The Living Planet Index (LPI) is a population-based indicator that combines data from individual populations of many vertebrate species to create a single index. The Global LPI for 2012 decreased by 28%. There are also indices that separate temperate and tropical species for marine and terrestrial species. The Red List Index is based on the IUCN Red List of Threatened Species and measures changes in conservation status over time and currently includes taxa that have been completely categorized: mammals, birds, amphibians and corals. The Global Wild Bird Index is another indicator that shows trends in population of wild bird groups on a regional scale from data collected in formal surveys. Challenges to these indices due to data availability are taxonomic gaps and the length of time of each index. The Biodiversity Indicators Partnership was established in 2006 to assist biodiversity indicator development, advancement and to increase the availability of indicators.

Importance of Biodiversity

Biodiversity is important for humans through ecosystem services and goods. Ecosystem services are broken down into: regulating services such as air and water purification, provisioning services (goods), such as fuel and food, cultural services and supporting services such as pollination and nutrient cycling.

REFERENCES

1. Diamond, Jared (1999). "Up to the Starting Line". Guns, Germs, and Steel. W. W. Norton. pp. 43–44. ISBN 0-393-31755-2.
2. Kunin, W.E.; Gaston, Kevin, eds. (31 December 1996). The Biology of Rarity: Causes and consequences of rare—common differences. ISBN 978-0412633805. Retrieved 26 May 2015.
3. Stearns, Beverly Peterson; Stearns, S. C.; Stearns, Stephen C. (2000). Watching, from the Edge of Extinction. Yale University Press. p. 1921. ISBN 978-0-300-08469-6. Retrieved 2014-12-27.
4. Novacek, Michael J. (8 November 2014). "Prehistory's Brilliant Future". New York Times. Retrieved 2014-12-25.
5. Newman, Mark (1997). "A model of mass extinction". Journal of Theoretical Biology. 189: 235–252. doi:10.1006/jtbi.1997.0508.
6. G. Miller; Scott Spoolman (2012). Environmental Science – Biodiversity Is a Crucial Part of the Earth's Natural Capital. Cengage Learning. p. 62. ISBN 1-133-70787-4. Retrieved 2014-12-27.
7. Mora, C.; Tittensor, D.P.; Adl, S.; Simpson, A.G.; Worm, B. (23 August 2011). "How many species are there on Earth and in the ocean?". PLOS Biology. 9: e1001127. doi:10.1371/journal.pbio.1001127. PMC 3160336 . PMID 21886479.
8. Sahney, S., Benton, M.J. and Ferry, P.A. (2010). "Links between global taxonomic diversity, ecological diversity and the expansion of vertebrates on land" (PDF).

Biology Letters. 6 (4): 544–547. doi:10.1098/rsbl.2009.1024. PMC 2936204 . PMID 20106856.

9. Species disappearing at an alarming rate, report says. MSNBC. Retrieved July 26, 2006.
10. Wilson, E.O., The Future of Life (2002) (ISBN 0-679-76811-4). See also: Leakey, Richard, The Sixth Extinction : Patterns of Life and the Future of Humankind, ISBN 0-385-46809-1
11. Davis, Paul and Kenrick, Paul. Fossil Plants. Smithsonian Books, Washington D.C. (2004). Morran, Robin, C.; A Natural History of Ferns. Timber Press (2004). ISBN 0-88192-667-1
12. See: Niles Eldredge, Time Frames: Rethinking of Darwinian Evolution and the Theory of Punctuated Equilibria, 1986, Heinemann ISBN 0-434-22610-6
13. Maas, Peter. "Extinct in the Wild" The Extinction Website. URL accessed January 26 2007.
14. Quince, C.; et al. "Deleting species from model food webs" (PDF). Archived from the original (PDF) on 2006-09-25. Retrieved 2007-02-15.
15. "Discovery" of the Coelacanth
16. 4 Extinct species that people still hope to rediscover – John R. Platt – Scientific American – February 21, 2013
17. Mills, L. Scott (2009-03-12). Conservation of Wildlife Populations: Demography, Genetics and Management. John Wiley & Sons. p. 13. ISBN 9781444308938.
18. Stearns, Beverly Peterson and Stephen C. (2000). "Preface". Watching, from the Edge of Extinction. Yale University Press. pp. x. ISBN 0-300-08469-2.
19. Raup, David M.; J. John Sepkoski Jr. (March 1982). "Mass extinctions in the marine fossil record". Science. 215 (4539): 1501–3. Bibcode:1982Sci...215.1501R. doi:10.1126/science.215.4539.1501. PMID 17788674.
20. "Population Bomb Author's Fix For Next Extinction: Educate Women". Scientific American. August 12, 2008.
21. "2004 Red List". IUCN Red List of Threatened Species. World Conservation Union. Retrieved September 20, 2006.[dead link]
22. Payne, J.L.; S. Finnegan (2007). "The effect of geographic range on extinction risk during background and mass extinction". Proc. Natl. Acad. Sci. 104 (25): 10506–11. Bibcode:2007PNAS..10410506P. doi:10.1073/pnas.0701257104. PMC 1890565 . PMID 17563357.
23. Mooney, H. A.; Cleland, E. E. (2001). "The evolutionary impact of invasive species". PNAS. 98 (10): 5446–5451. Bibcode:2001PNAS...98.5446M. doi:10.1073/pnas.091093398. PMC 33232 . PMID 11344292.
24. Glossary: definitions from the following publication: Aubry, C., R. Shoal and V. Erickson. 2005. Grass cultivars: their origins, development, and use on national forests and grasslands in the Pacific Northwest. USDA Forest Service. 44 pages, plus appendices.; Native Seed Network (NSN), Institute for Applied Ecology, 563 SW Jefferson Ave, Corvallis, OR 97333, USA

25. "Australia's state of the forests report" (PDF). 2003. p. 107.

26. Rhymer, J. M.; Simberloff, D. (November 1996). "Extinction by Hybridization and Introgression". Annual Review of Ecology and Systematics. Annual Reviews. 27: 83–109. doi:10.1146/annurev.ecolsys.27.1.83. Introduced species, in turn, are seen as competing with or preying on native species or destroying their habitat. Introduces species (or subspecies), however, can generate another kind of extinction, a genetic extinction by hybridization and introgression with native flora and fauna.

27. Potts, Brad M. (September 2001). "Genetic pollution from farm forestry using eucalypt species and hybrids : a report for the RIRDC/L&WA/FWPRDC Joint Venture Agroforestry Program". Robert C. Barbour, Andrew B. Hingston. Australian Government, Rural Industrial Research and Development Corporation. ISBN 0-642-58336-6.

28. "GENETIC DIVERSITY" (PDF). 2003. p. 104. Retrieved 2010-05-30. In other words, greater genetic diversity can offer greater resilience. In order to maintain the capacity of our forests to adapt to future changes, therefore, genetic diversity must be preserved

29. Lindenmayer, D. B.; Hobbs, R. J.; Salt, D. (2003-01-06). "Plantation forests and biodiversity conservation" (PDF). Australian Forestry. 66 (1): 64. there may be genetic invasion from pollen dispersal and subsequent hybridisation between eucalypt tree species used to establish plantations and eucalypts endemic to an area (Potts et al. 2001). This may, in turn, alter natural patterns of genetic variability

30. Clover, Charles (2004). The End of the Line: How overfishing is changing the world and what we eat. London: Ebury Press. ISBN 0-09-189780-7.

31. Lee, Anita. "The Pleistocene Overkill Hypothesis." University of California at Berkeley Geography Program.'.' Retrieved January 11, 2007.

32. Koh, Lian Pih. Science, Vol 305, Issue 5690, 1632–1634, 10 September 2004.

33. Dunn, Robert; Nyeema Harris; Robert Colwell; Lian Pin Koh; Navjot Sodhi (2009). "Proceedings of the Royal Society". The sixth mass coextinction: are most endangered species parasites and mutualists?. The Royal Society. Retrieved April 20, 2011.

34. Sahney, S., Benton, M.J. & Falcon-Lang, H.J. (2010). "Rainforest collapse triggered Pennsylvanian tetrapod diversification in Euramerica" (PDF). Geology. 38 (12): 1079–1082. Bibcode:2010Geo....38.1079S. doi:10.1130/G31182.1.

35. Thomas, C. D.; et al. (2004-01-08). "Extinction risk from climate change". Nature. 427 (6970): 145–148. Bibcode:2004Natur.427..145T. doi:10.1038/nature02121. PMID 14712274. Retrieved 2010-05-28. minimal climate-warming scenarios produce lower projections of species committed to extinction (approx18%) (Letter to Nature received 10 September 2003.)

36. Bhattacharya, Shaoni (7 January 2004). "Global warming threatens millions of species". New Scientist. Retrieved 2010-05-28. the effects of climate change should be considered as great a threat to biodiversity as the "Big Three" –

habitat destruction, invasions by alien species and overexploitation by humans.

37. Handwerk, Brian, and Brian Hendwerk. "Global Warming Could Cause Mass Extinctions by 2050, Study Says." National Geographic News (Apr. 2006): n. pag. www.nationalgeographic.com. Web. 12 Oct. 2009.
38. Sahney, S.; Benton, M.J. (2008). "Recovery from the most profound mass extinction of all time" (PDF). Proceedings of the Royal Society: Biological. 275 (1636): 759–65. doi:10.1098/rspb.2007.1370. PMC 2596898 . PMID 18198148.
39. Benton M J (2005). When Life Nearly Died: The Greatest Mass Extinction of All Time. Thames & Hudson. ISBN 978-0-500-28573-2.
40. American Museum of Natural History. "National Survey Reveals Biodiversity Crisis – Scientific Experts Believe We are in the Midst of the Fastest Mass Extinction in Earth's History". Retrieved September 20, 2006.
41. Wilson, E. O. (April 30, 2012). ""E. O. Wilson wants to know why you're not protesting in the streets"". Grist (Interview). Interview with Lisa Hymas. Retrieved January 16, 2014. E. O. Wilson repeats his estimation in 2012.
42. J.H.Lawton and R.M.May, Extinction rates, Oxford University Press, Oxford, UK
43. De Vos; et al. (2014). "Estimating the Normal Background Rate of Species Extinction". Conservation Biology. 29: 452–462. doi:10.1111/cobi.12380.
44. Viney, Mike. "Extinction Part 2 of 5". Colorado State University. Retrieved September 12, 2006.
45. Academy of Natural Sciences, "Fossils and Extinction" <http://www.ansp.org/museum/jefferson/otherPages/extinction.php> and U.C. Berkeley "History of Evolutionary Thought — Extinction" http://evolution.berkeley.edu/evosite/history/extinction.shtml.
46. Koerner, Lisbet (1999). "God's Endless Larder". Linnaeus: Nature and Nation. Harvard University Press. p. 85. ISBN 0-674-00565-1.
47. Ideas: A History from Fire to Freud (Peter Watson Weidenfeld & Nicolson ISBN 0-297-60726-X)
48. Watson, p.16
49. Robert Chambers, 1844, Vestiges of the Natural History of Creation, 1994 reprint: University of Chicago Press ISBN 0-226-10073-1
50. C. Cormack Gates; Curtis H. Freese; Peter J.P. Gogan; Mandy Kotzman. American bison: status survey and conservation guidelines 2010. IUCN. p. 15. ISBN 978-2-8317-1149-2. Retrieved 6 November 2011.
51. Walsh, Bruce. Extinction. Bioscience at University of Arizona. Retrieved July 26, 2006.
52. Committee on Recently Extinct Organisms. "Why Care About Species That Have Gone Extinct?". Retrieved July 30, 2006.
53. International Programme on Chemical Safety (1989). "DDT and its Derivatives – Environmental Aspects". Environmental Health Criteria 83. Retrieved September 20, 2006.

54. Diamond, Jared (2005). "A Tale of Two Farms". Collapse. Penguin. pp. 15–17. ISBN 0-670-03337-5.
55. Drewry, Rachel. "Ecotourism: Can it save the orangutans?" Inside Indonesia. Retrieved January 26, 2007.
56. The Wildlands Project. Retrieved January 26, 2007.
57. Alliance for Zero Extinctions. Retrieved January 26, 2007.
58. Ehrlich, Anne (1981). Extinction: The Causes and Consequences of the Disappearance of Species. Random House, New York. ISBN 0-394-51312-6.
59. "Smallpox". WHO Factsheet. Archived from the original on 2007-09-21.
60. Dennis Normile (2008). "Driven to Extinction". Science. 319 (5870): 1606–1609. doi:10.1126/science.319.5870.1606. PMID 18356500. Retrieved 2009-03-28.
61. "Polio cases in the world in 2015". The Global Polio Eradication Initiative. Retrieved 17 February 2016.
62. "This Species is Close to Extinction and That's a Good Thing". Time. 23 January 2015. Retrieved 17 February 2016.
63. Olivia Judson (September 25, 2003). "A Bug's Death". The New York Times. Retrieved 17 February 2016.
64. A. Zitner (2000-12-24). "Cloned Goat Would Revive Extinct Line". Los Angeles Times. Retrieved 2010-05-17.
65. Nicholas wade (2008-11-19). "Regenerating a Mammoth for $10 Million". The New York Times. Retrieved 2010-05-17. The cell could be converted into an embryo and brought to term by an elephant, a project he estimated would cost some $10 million. "This is something that could work, though it will be tedious and expensive,"
66. Steve Connor (2009-02-02). "Cloned goat dies after attempt to bring species back from extinction". The Independent. London. Retrieved 2010-05-17.
67. Gray, Richard; Dobson, Roger (31 Jan 2009). "Extinct ibex is resurrected by cloning". Telegraph. London.
68. McKinney 1997, p. 110
69. Stearns & Stearns 1999, p. x
70. Novacek, Michael J. (8 November 2014). "Prehistory's Brilliant Future". The New York Times. New York: The New York Times Company. ISSN 0362-4331. Retrieved 2014-12-25.
71. http://www.pnas.org/content/early/2016/04/26/1521291113.full
72. "Catalogue of Life: 2016 Annual Checklist". 2016. Retrieved 2016-11-06.
73. Mora, Camilo; Tittensor, Derek P.; Adl, Sina; et al. (23 August 2011). "How Many Species Are There on Earth and in the Ocean?". PLOS Biology. San Francisco, CA: PLOS. 9 (8): e1001127. doi:10.1371/journal.pbio.1001127. ISSN 1545-7885. PMC 3160336 . PMID 21886479.
74. Staff (2 May 2016). "Researchers find that Earth may be home to 1 trillion

species". National Science Foundation. Retrieved 6 May 2016.

75. Nuwer, Rachel (18 July 2015). "Counting All the DNA on Earth". The New York Times. New York: The New York Times Company. ISSN 0362-4331. Retrieved 2015-07-18.
76. "The Biosphere: Diversity of Life". Aspen Global Change Institute. Basalt, CO. Retrieved 2015-07-19.
77. Chapman, A. D. (2009). Numbers of Living Species in Australia and the World (PDF) (2nd ed.). Canberra: Australian Biological Resources Study. pp. 1–80. ISBN 978 0 642 56861 8.
78. Costello, Mark; Robert May; Nigel Stork (25 January 2013). "Can we name Earth's species before they go extinct?". Science. 339 (6118): 413–416. doi:10.1126/science.1230318. PMID 23349283.
79. Sweetlove, Lee. "Number of species on Earth tagged at 8.7 million". Nature. Macmillan Publishers Limited. Retrieved 18 July 2014.
80. Donald R. Prothero (2013), Bringing Fossils to Life: An Introduction to Paleobiology (3rd ed.), Columbia University Press, p. 21
81. Sahney, S., Benton, M.J. & Falcon-Lang, H.J. (2010). "Rainforest collapse triggered Pennsylvanian tetrapod diversification in Euramerica" (PDF). Geology. 38 (12): 1079–1082. doi:10.1130/G31182.1.
82. Barnosky, A. D., et al. (2011). "Has the Earth's sixth mass extinction already arrived?". Nature. 471 (7336): 51–57. doi:10.1038/nature09678.
83. Pereira, HM. "Global Biodiversity Change: The Bad, the Good, and the Unknown". Annual Review of Environment and Resources.
84. Sahney, S.; Benton, M.J. (2008). "Recovery from the most profound mass extinction of all time" (PDF). Proceedings of the Royal Society: Biological. 275 (1636): 759–65. doi:10.1098/rspb.2007.1370. PMC 2596898 . PMID 18198148.
85. Sahney, S., Benton, M.J. and Ferry, P.A. (2010). "Links between global taxonomic diversity, ecological diversity and the expansion of vertebrates on land" (PDF). Biology Letters. 6 (4): 544–547. doi:10.1098/rsbl.2009.1024. PMC 2936204 . PMID 20106856.
86. Raup. D.M. (1986). "Biological extinction in earth history". Science. 231 (4745): 1528–1533. doi:10.1126/science.11542058. PMID 11542058.
87. IISE (2010). SOS 2009: State of Observed Species (PDF). Arizona State University: International Institute for Species Exploration. pp. 1–8.
88. IISE (2011). SOS 2010: State of Observed Species (PDF). Arizona State University: International Institute for Species Exploration. pp. 1–10.
89. IISE (2012). SOS 2011: State of Observed Species (PDF). Arizona State University: International Institute for Species Exploration. pp. 1–14.
90. Encyclopedia Smithsonian: Numbers of Insects
91. Proceedings of the National Academy of Sciences, Census of Marine Life (CoML) BBC News

92. David L. Hawksworth, "The magnitude of fungal diversity: the 1•5 million species estimate revisited" Mycological Research (2001), 105: 1422-1432 Cambridge University Press Abstract
93. Acari at University of Michigan Museum of Zoology Web Page
94. Pawlowski, J. et al. (2012). CBOL Protist Working Group: Barcoding Eukaryotic Richness beyond the Animal, Plant, and Fungal Kingdoms. PLoS Biol 10(11): e1001419. doi:10.1371/journal.pbio.1001419, .
95. Adl, S. M. et al. (2007). Diversity, nomenclature, and taxonomy of protists. Systematic Biology 56(4), 684-689, .
96. Erwin, Terry L. (March 1982). The Coleopterists Society, ed. "Tropical Forests: Their Richness in Coleoptera and Other Arthropod Species". The Coleopterists Bulletin. 36 (1): 74–75. ISSN 0010-065X. JSTOR 4007977.
97. Pullin, Andrew (2002). Conservation Biology. Cambridge: Cambridge University Press. ISBN 9780521644822. Retrieved December 24, 2013.
98. "Researchers find that Earth may be home to 1 trillion species". NSF. 2 May 2016. Retrieved 6 May 2016.
99. . Zoological Society of London http://www.zsl.org/science/research-projects/indicators-assessments/index,134,ZI.html. Missing or empty |title= (help)
100. "Trends in the status of biodiversity". IUCN. Retrieved 25 November 2013.
101. "Global Wild Bird Index". Biodiversity Indicators Partnership.
102. De Groot, R.S.; et al. (2002). "A typology for the classification, and description and valuation of ecosystem functions, goods and services". Ecological Economics. 41 (3): 393–408. doi:10.1016/s0921-8009(02)00089-7.
103. Berlocher 1998, p. 3
104. Cook, Orator F. (March 30, 1906). "Factors of species-formation". Science. Washington, D.C.: American Association for the Advancement of Science. 23 (587): 506–507. doi:10.1126/science.23.587.506. ISSN 0036-8075. PMID 17789700.
105. Cook, Orator F. (November 1908). "Evolution Without Isolation". The American Naturalist. Chicago, IL: University of Chicago Press on behalf of the American Society of Naturalists. 42 (503): 727–731. doi:10.1086/279001. ISSN 0003-0147.
106. Via, Sara (June 16, 2009). "Natural selection in action during speciation" (PDF). Proc. Natl. Acad. Sci. U.S.A. Washington, D.C.: National Academy of Sciences. 106 (Suppl 1): 9939–9946. doi:10.1073/pnas.0901397106. ISSN 0027-8424. PMC 2702801 . PMID 19528641.
107. Baker, Jason M. (June 2005). "Adaptive speciation: The role of natural selection in mechanisms of geographic and non-geographic speciation". Studies in History and Philosophy of Science Part C: Studies in History and Philosophy of Biological and Biomedical Sciences. Amsterdam, the Netherlands: Elsevier. 36 (2): 303–326. doi:10.1016/j.shpsc.2005.03.005. ISSN 1369-8486. PMID 19260194.
108. Kingsley, David M. (January 2009). "Diversity Revealed: From Atoms to Traits". Scientific American. Stuttgart: Georg von Holtzbrinck Publishing Group. 300

(1): 52–59. doi:10.1038/scientificamerican0109-52. ISSN 0036-8733.

109. Sulloway, Frank J. (September 30, 1982). "The Beagle collections of Darwin's finches (Geospizinae)". Bulletin of the British Museum (Natural History). Zoology. London: British Museum (Natural History). 43 (2): 49–58. ISSN 0007-1498.
110. Mayr 1992, pp. 21–53
111. Endler 1977
112. Tarkhnishvili, David; Murtskhvaladze, Marine; Gavashelishvili, Alexander (August 2013). "Speciation in Caucasian lizards: climatic dissimilarity of the habitats is more important than isolation time". Biological Journal of the Linnean Society. Hoboken, NJ: Wiley-Blackwell on behalf of the Linnean Society of London. 109 (4): 876–892. doi:10.1111/bij.12092. ISSN 0024-4074.
113. Feder, Jeffrey L.; Xianfa Xie; Rull, Juan; et al. (May 3, 2005). "Mayr, Dobzhansky, and Bush and the complexities of sympatric speciation inRhagoletis" (PDF). Proc. Natl. Acad. Sci. U.S.A. Washington, D.C.: National Academy of Sciences. 102 (Suppl 1): 6573–6580. doi:10.1073/pnas.0502099102. ISSN 0027-8424. PMC 1131876 . PMID 15851672.
114. Berlocher, Stewart H.; Feder, Jeffrey L. (January 2002). "Sympatric Speciation in Phytophagous Insects: Moving Beyond Controversy?". Annual Review of Entomology. Palo Alto, CA: Annual Reviews. 47: 773–815. doi:10.1146/annurev.ento.47.091201.145312. ISSN 0066-4170. PMID 11729091.
115. Daniel I. Bolnick and Benjamin M. Fitzpatrick (2007), "Sympatric Speciation: Models and Empirical Evidence", Annual Review of Ecology, Evolution, and Systematics, 38: 459–487, doi:10.1146/annurev.ecolsys.38.091206.095804
116. Machado, Heather E.; Pollen, Alexander A.; Hofmann, Hans A.; et al. (December 2009). "Interspecific profiling of gene expression informed by comparative genomic hybridization: A review and a novel approach in African cichlid fishes". Integrative and Comparative Biology. Oxford, UK: Oxford University Press for the Society for Integrative and Comparative Biology. 49 (6): 644–659. doi:10.1093/icb/icp080. ISSN 1557-7023. PMID 21665847.
117. Fan, Shaohua; Elmer, Kathryn R.; Meyer, Axel (February 5, 2012). "Genomics of adaptation and speciation in cichlid fishes: recent advances and analyses in African and Neotropical lineages". Philosophical Transactions of the Royal Society B. London: Royal Society. 367 (1587): 385–394. doi:10.1098/rstb.2011.0247. ISSN 0962-8436. PMC 3233715 . PMID 22201168.
118. Niemiller, Matthew L.; Fitzpatrick, Benjamin M.; Miller, Brian T. (May 2008). "Recent divergence with gene flow in Tennessee cave salamanders (Plethodontidae: Gyrinophilus) inferred from gene genealogies". Molecular Ecology. Hoboken, NJ: Wiley-Blackwell. 17 (9): 2258–2275. doi:10.1111/j.1365-294X.2008.03750.x. ISSN 0962-1083. PMID 18410292.
119. Martens, Koen (May 1997). "Speciation in ancient lakes". Trends in Ecology & Evolution. 12 (5): 177–182. doi:10.1016/S0169-5347(97)01039-2.
120. *Joly, E (9 December 2011). "The existence of species rests on a metastable equilibrium between inbreeding and outbreeding. An essay on the close relationship*

between speciation, inbreeding and recessive mutations.”. Biology direct. 6: 62. doi:10.1186/1745-6150-6-62. PMC 3275546 . PMID 22152499.

121. Feder, Jeffrey L.; Roethele, Joseph B.; Filchak, Kenneth; et al. (March 2003). “Evidence for inversion polymorphism related to sympatric host race formation in the apple maggot fly, Rhagoletis pomonella”. Genetics. Bethesda, MD: Genetics Society of America. 163 (3): 939–953. ISSN 0016-6731. PMC 1462491 . PMID 12663534. Retrieved 2015-09-07.

122. Berlocher, Stewart H.; Bush, Guy L. (June 1982). “An electrophoretic analysis of Rhagoletis (Diptera: Tephritidae) phylogeny”. Systematic Zoology. Taylor & Francis, Ltd., on behalf of the Society of Systematic Biologists. 31 (2): 136–155. doi:10.2307/2413033. JSTOR 2413033.

123. Ridley, Mark. “Speciation - What is the role of reinforcement in speciation?”. Retrieved 2015-09-07. Adapted from Evolution (2004), 3rd edition (Malden, MA: Blackwell Publishing), ISBN 978-1-4051-0345-9.

124. Ollerton, Jeff (September 2005). “Speciation: Flowering time and the Wallace Effect” (PDF). Heridity. London: Nature Publishing Group for The Genetics Society. 95 (3): 181–182. doi:10.1038/sj.hdy.6800718. ISSN 0018-067X. PMID 16077739. Retrieved 2015-09-07.

125. Howard D. Rundle and Patrik Nosil (2005), “Ecological speciation”, Ecology Letters, 8: 336–352, doi:10.1111/j.1461-0248.2004.00715.x

126. Dolph Schluter (2001), “Ecology and the origin of species”, Trends in Ecology and Evolution, 16 (7)

127. Jeffrey S. McKinnon; et al. (2004), “Evidence for ecology’s role in speciation”, Nature, 429: 294–298, doi:10.1038/nature02556

128. Dolph Schluter (2009), “Evidence for Ecological Speciation and Its Alternative”, Science, 326: 737–740

129. Panhuis, Tami M.; Butlin, Roger; Zuk, Marlene; et al. (July 2001). “Sexual selection and speciation” (PDF). Trends in Ecology & Evolution. Cambridge, MA: Cell Press. 16 (7): 364–371. doi:10.1016/s0169-5347(01)02160-7. ISSN 0169-5347. PMID 11403869.

130. Darwin, Charles; A. R. Wallace (1858). “On the Tendency of Species to form Varieties; and on the Perpetuation of Varieties and Species by Natural Means of Selection” (PDF). Journal of the Proceedings of the Linnean Society of London. Zoology. 3: 46–50. doi:10.1111/j.1096-3642.1858.tb02500.x.

131. Darwin, Charles (1859). On the Origin of Species (1st edition). Chapter 4, page 89. http://darwin-online.org.uk/content/frameset?viewtype=side&itemID=F373&pageseq=12

132. Eberhard, W. G. (1985). Sexual Selection and Animal Genitalia. Harvard University Press, Cambridge, Mass.

133. Gould, Stephen Jay (1980). “A Quahog is a Quahog.”. in: The Panda’s thumb. More reflections in natural history. New York: W. W. Norton & Company. pp. 204–213. ISBN 0-393-30023-4.

134. Maynard Smith 1989, pp. 275–280

135. Darwin 1859

136. Bernstein, Harris; Byerly, Henry C.; Hopf, Frederic A.; et al. (December 21, 1985). "Sex and the emergence of species". Journal of Theoretical Biology. Amsterdam, the Netherlands: Elsevier. 117 (4): 665–690. doi:10.1016/S0022-5193(85)80246-0. ISSN 0022-5193. PMID 4094459.

137. Koeslag, Johan H. (May 10, 1990). "Koinophilia groups sexual creatures into species, promotes stasis, and stabilizes social behaviour". Journal of Theoretical Biology. Amsterdam, the Netherlands: Elsevier. 144 (1): 15–35. doi:10.1016/s0022-5193(05)80297-8. ISSN 0022-5193. PMID 2200930.

138. Koeslag, Johan H. (December 21, 1995). "On the Engine of Speciation". Journal of Theoretical Biology. Amsterdam, the Netherlands: Elsevier. 177 (4): 401–409. doi:10.1006/jtbi.1995.0256. ISSN 0022-5193.

139. Poelstra, Jelmer W.; Vijay, Nagarjun; Bossu, Christen M.; et al. (June 20, 2014). "The genomic landscape underlying phenotypic integrity in the face of gene flow in crows". Science. Washington, D.C.: American Association for the Advancement of Science. 344 (6190): 1410–1414. doi:10.1126/science.1253226. ISSN 0036-8075. PMID 24948738. The Phenotypic Differences between Carrion and Hooded Crows across the Hybridization Zone in Europe are Unlikely to be due to Assortative Mating. — Commentary by Mazhuvancherry K. Unnikrishnan and H. S. Akhila

140. Miller 2013, pp. 177, 395–396

141. Nowak 1999

142. Hiendleder, Stefan; Kaupe, Bernhard; Wassmuth, Rudolf; et al. (May 7, 2002). "Molecular analysis of wild and domestic sheep questions current nomenclature and provides evidence for domestication from two different subspecies". Proceedings of the Royal Society B. London: Royal Society. 269 (1494): 893–904. doi:10.1098/rspb.2002.1975. ISSN 0962-8452. PMC 1690972 . PMID 12028771.

143. Rice, William R.; Salt, George W. (June 1988). "Speciation Via Disruptive Selection on Habitat Preference: Experimental Evidence". The American Naturalist. Chicago, IL: University of Chicago Press on behalf of the American Society of Naturalists. 131 (6): 911–917. doi:10.1086/284831. ISSN 0003-0147.

144. Rice, William R.; Hostert, Ellen E. (December 1993). "Laboratory Experiments on Speciation: What Have We Learned in 40 Years?". Evolution. Hoboken, NJ: John Wiley & Sons for the Society for the Study of Evolution. 47 (6): 1637–1653. doi:10.2307/2410209. ISSN 0014-3820. JSTOR 2410209.

145. Gavrilets, Sergey (October 2003). "Perspective: Models of Speciation: What Have We Learned in 40 Years?". Evolution. Hoboken, NJ: John Wiley & Sons for the Society for the Study of Evolution. 57 (10): 2197–2215. doi:10.1554/02-727. ISSN 0014-3820. PMID 14628909.

146. Dodd, Diane M. B. (September 1989). "Reproductive Isolation as a Consequence of Adaptive Divergence in Drosophila pseudoobscura". Evolution. Hoboken, NJ: John Wiley & Sons for the Society for the Study of Evolution. 43 (6):

1308–1311. doi:10.2307/2409365. ISSN 0014-3820. JSTOR 2409365.

147. Kirkpatrick, Mark; Ravigné, Virginie (March 2002). "Speciation by Natural and Sexual Selection: Models and Experiments". The American Naturalist. Chicago, IL: University of Chicago Press on behalf of the American Society of Naturalists. 159 (S3): S22–S35. doi:10.1086/338370. ISSN 0003-0147. PMID 18707367.

148. Koukou, Katerina; Pavlikaki, Haris; Kilias, George; et al. (January 2006). "Influence of Antibiotic Treatment and Wolbachia Curing on Sexual Isolation Among Drosophila melanogaster Cage Populations". Evolution. Hoboken, NJ: John Wiley & Sons for the Society for the Study of Evolution. 60 (1): 87–96. doi:10.1554/05-374.1. ISSN 0014-3820. PMID 16568634.

149. Symons 1979

150. Langlois, Judith H.; Roggman, Lori A. (March 1990). "Attractive Faces Are Only Average". Psychological Science. Thousand Oaks, CA: Sage Publications. 1 (2): 115–121. doi:10.1111/j.1467-9280.1990.tb00079.x. ISSN 0956-7976.

151. Phadnis, Nitin; Orr, H. Allen (January 16, 2009). "A Single Gene Causes Both Male Sterility and Segregation Distortion in DrosophilaHybrids". Science. Washington, D.C.: American Association for the Advancement of Science. 323 (5912): 376–379. doi:10.1126/science.1163934. ISSN 0036-8075. PMC 2628965 . PMID 19074311.

152. Ramsey, Justin; Schemske, Douglas W. (November 1998). "Pathways, Mechanisms, and Rates of Polyploid Formation in Flowering Plants". Annual Review of Ecology and Systematics. Palo Alto, CA: Annual Reviews. 29: 467–501. doi:10.1146/annurev.ecolsys.29.1.467. ISSN 1545-2069.

153. Otto, Sarah P.; Whitton, Jeannette (December 2000). "Polyploid Incidence and Evolution". Annual Review of Genetics. Palo Alto, CA: Annual Reviews. 34: 401–437. doi:10.1146/annurev.genet.34.1.401. ISSN 0066-4197. PMID 11092833.

154. Comai, Luca (November 2005). "The advantages and disadvantages of being polyploid". Nature Reviews Genetics. London: Nature Publishing Group. 6 (11): 836–846. doi:10.1038/nrg1711. ISSN 1471-0056. PMID 16304599.

155. Wendel, Jonathan F. (January 2000). "Genome evolution in polyploids". Plant Molecular Biology. Kluwer Academic Publishers. 42 (1): 225–249. doi:10.1023/A:1006392424384. ISSN 0167-4412. PMID 10688139.

156. Sémon, Marie; Wolfe, Kenneth H. (December 2007). "Consequences of genome duplication". Current Opinion in Genetics & Development. Amsterdam, the Netherlands: Elsevier. 17 (6): 505–512. doi:10.1016/j.gde.2007.09.007. ISSN 0959-437X. PMID 18006297.

157. Soltis, Pamela S.; Soltis, Douglas E. (June 20, 2000). "The role of genetic and genomic attributes in the success of polyploids". Proc. Natl. Acad. Sci. U.S.A. Washington, D.C.: National Academy of Sciences. 97 (13): 7051–7057. doi:10.1073/pnas.97.13.7051. ISSN 0027-8424. PMC 34383 . PMID 10860970.

158. Mavarez, Jesús; Salazar, Camilo A.; Bermingham, Eldredge; et al. (June 15, 2006). "Speciation by hybridization in Heliconius butterflies". Nature. London: Nature Publishing Group. 441 (7095): 868–871. doi:10.1038/nature04738. ISSN

0028-0836. PMID 16778888.

159. Sherwood, Jonathan (September 8, 2006). "Genetic Surprise Confirms Neglected 70-Year-Old Evolutionary Theory" (Press release). University of Rochester. Retrieved 2015-09-10.

160. Masly, John P.; Jones, Corbin D.; Mohamed, A. F. Noor; et al. (September 8, 2006). "Gene Transposition as a Cause of Hybrid Sterility in Drosophila". Science. Washington, D.C.: American Association for the Advancement of Science. 313 (5792): 1448–1450. doi:10.1126/science.1128721. ISSN 0036-8075. PMID 16960009.

161. Minkel, J. R. (September 8, 2006). "Wandering Fly Gene Supports New Model of Speciation". Scientific American. Stuttgart: Georg von Holtzbrinck Publishing Group. ISSN 0036-8733. Retrieved 2015-09-11.

162. Mayr 1982, p. 273

163. Hopf, Frederic A.; Hopf, F. W. (February 1985). "The role of the Allee effect in species packing". Theoretical Population Biology. Amsterdam, the Netherlands: Elsevier. 27 (1): 27–50. doi:10.1016/0040-5809(85)90014-0. ISSN 0040-5809.

164. Bernstein & Bernstein 1991

165. Michod 1995

166. Michod 1999

167. Hockey, Dean & Ryan 2005, pp. 176, 193

168. Mayr 1988

169. Williams 1992, p. 118

170. Maynard Smith, John (December 1983). "The Genetics of Stasis and Punctuation". Annual Review of Genetics. Palo Alto, CA: Annual Reviews. 17: 11–25. doi:10.1146/annurev.ge.17.120183.000303. ISSN 0066-4197. PMID 6364957.

171. Clapham, Tutin & Warburg 1952

172. Grant 1971

173. Gould, Stephen Jay; Eldredge, Niles (Spring 1977). "Punctuated equilibria: the tempo and mode of evolution reconsidered" (PDF). Paleobiology. Boulder, CO: Paleontological Society. 3 (2): 115–151. doi:10.1017/s0094837300005224. ISSN 0094-8373. JSTOR 2400177. Archived from the original (PDF) on 2014-06-24. Retrieved 2015-09-15.

174. Laws 2010, pp. 210–215

175. Williams 1992, chpt. 9

176. Eldredge & Gould 1972, chpt. 5

177. Mayr 1954, pp. 157–180

178. Maynard Smith 1989, p. 281

179. Gould 1980, pt. 4, chpt. 18

180. Williams 1974

181. Maynard Smith, John (March 14, 1964). "Group Selection and Kin Selection". Nature. London: Nature Publishing Group. 201 (4924): 1145–1147. doi:10.1038/2011145a0. ISSN 0028-0836.

182. Dawkins 1995, chpt. 4

183. Dawkins, Richard (December 1994). "Burying the Vehicle". Behavioural and Brain Sciences. Cambridge, UK: Cambridge University Press. 17 (4): 616–617. doi:10.1017/S0140525X00036207. ISSN 0140-525X. Archived from the original on 2006-09-15. Retrieved 2015-09-15. "Remarks on an earlier article by [Elliot] Sober [sic] and David Sloan Wilson, who made a more extended argument in their recent book Unto Others : The Evolution and Psychology of Unselfish Behavior"

184. Dennett, Daniel C. (December 1994). "E Pluribus Unum?". Behavioural and Brain Sciences. Cambridge, UK: Cambridge University Press. 17 (4): 617–618. doi:10.1017/S0140525X00036219. ISSN 0140-525X. Archived from the original on 2007-12-27."Commentary on Wilson & Sober: Group Selection."

185. Pinker, Steven (June 18, 2012). "The False Allure of Group Selection". edge. org. Edge Foundation, Inc. Retrieved 2015-09-15.

186. Campbell 1990, pp. 450–451, 487–490, 499–501

187. Ayala 1982, pp. 73–83, 182–190, 198–215

188. McCarthy & Rubidge 2005

189. E.g. Julia Whitty (2007-04-30). "Animal Extinction - the greatest threat to mankind: By the end of the century half of all species will be extinct. Does that matter?". The Independent. Archived from the original on 2015-08-06. Retrieved 2010-05-20. By the most conservative measure - based on the last century's recorded extinctions - the current rate of extinction is 100 times the background rate.

190. Thackeray, J. Francis (1990). "Rates of Extinction in Marine Invertebrates: Further Comparison Between Background and Mass Extinctions". Paleobiology. Paleontological Society. 16 (1): 22–4. ISSN 1938-5331. JSTOR 2400930 – via JSTOR. (registration required (help)).

191. N.L. Gilinsky (1994). "Volatility and the Phanerozoic decline of background extinction intensity". Paleobiology. 20 (4): 445–458. JSTOR 2401228 – via JSTOR. (registration required (help)).

192. Raymond, H, Ward, P: "Hypoxia, Global Warming, and Terrestrial Late Permian Extinctions" Page 389–401. Science 15, 2005. http://www.sciencemag.org/cgi/content/full/308/5720/398

193. American Museum of Natural History, 1998. http://www.amnh.org/science/biodiversity/extinction/Intro/OngoingProcess.html

194. Pimm, S.: "The Extinction Puzzle", Project Syndicate, 2007. http://www.project-syndicate.org/commentary/pimm1

195. May, R. Lawton, J. Stork, N: "Assessing Extinction Rates" Oxford University Press, 1995.

196. Lawton, John H.; May, Robert McCredie (1995-01-01). Extinction Rates. Oxford University Press. ISBN 9780198548294.

CHAPTER 3

Agricultural Biodiversity

Chapter Contents

- **Introduction**
- **Overexploitation**
- **Overgrazing**

Agricultural biodiversity provides not only food and income but also raw materials for clothing, shelter, medicines, breeding new varieties, and performs other services such as maintenance of soil fertility and biota, and soil and water conservation. All of which are essential to human survival. Nearly one third of the world's land area is used for food production.

INTRODUCTION

Agricultural biodiversity is a sub-set of general biodiversity. It includes all forms of life directly relevant to agriculture: rare seed varieties and animal breeds (farm biodiversity), but also many other organisms such as soil fauna, weeds, pests, predators, and all of the native plants and animals (wild biodiversity) existing on and flowing through the farm. However, most attention in this field is given to crop varieties and to crop wild relatives. Cultivated varieties can be broadly classified into "modern varieties" and "farmer's or traditional varieties". Modern varieties are the outcome of formal breeding and are often characterized as 'high yielding'. For example, the short straw wheat and rice varieties of the Green Revolution. In contrast, farmer's varieties (also known as landraces) are the product of (breeding and) selection carried out by farmers. Together, these varieties represent high levels of genetic diversity and are therefore the focus of most crop genetic resources conservation efforts. Agricultural biodiversity will also be absolutely essential to cope with the predicted impacts of climate change, not simply as a source of traits but as the underpinnings of more resilient farm ecosystems. Agricultural biodiversity is the basis of our agricultural food chain, developed and safeguarded by farmers, livestock breeders, forest workers, fishermen and indigenous peoples throughout the world. The use of agricultural biodiversity (as opposed to non diverse production methods) can contribute to food security and livelihood security.

Unusual strains of maize are collected to increase the crop diversity when selectively breeding domestic corn.

Scope

Although the term *agricultural biodiversity* is relatively new - it has come into wide use in recent years as evidenced by bibliographic references - the concept itself is quite old. It is the result of the careful selection and inventive developments of farmers, herders and fishers over millennia. Agricultural biodiversity is a vital sub-set of biodiversity. It is a use of life, i.e. ancillary biotechnologies, by Mankind whose food and livelihood security depend on the sustained management of those diverse biological resources that are important for food and agriculture. As for everything, agricultural biodiversity can be used, not used, misused, and even abused. Agricultural biodiversity includes:

- Domesticated crop and 'wild' plants (called: crop wild relatives), including woody perennials (see: forest genetic resources) and aquatic plants (used for food and other natural resources based products), domestic and wild animals (used for food, fibre, milk, hides, furs, power, organic fertilizer), fish and other aquatic animals, within field, forest, rangeland, and aquatic ecosystems
- Domesticated livestock species and their wild relatives (http://www.fao.org/dad-is)
- Non-harvested species within production agroecosystems that support food provision, including soil micro-biota, pollinators, and so on

- Non-harvested species in the wider environment that support food production agroecosystems (agricultural, pastoral, forest, and aquatic ecosystems)

However, agricultural biodiversity, sometimes called Agrobiodiversity, *encompasses the variety and variability of* animals, plants, and micro-organisms which are necessary to sustain key functions of the agroecosystem, its structure and processes for, and in support of, food production and food security". It further "comprises genetic, population, species, community, ecosystem, and landscape components, and human interactions with all these."

Aquatic diversity is also an important component of agricultural biodiversity. The conservation and sustainable use of local aquatic ecosystems, ponds, rivers, coastal commons by artisanal fisherfolk and smallholder farmers is important to the survival of both humans and the environment. Since aquatic organisms, including fish, provide much of our food supply as well as underpinning the income of coastal peoples, it is critical that fisherfolk and smallholder farmers have genetic reserves and sustainable ecosystems to draw upon as aquaculture and marine fisheries management continue to evolve.

Genetic Erosion in Agricultural and Livestock Biodiversity

Genetic erosion in agricultural and livestock biodiversity is the loss of genetic diversity, including the loss of individual genes, and the loss of particular combinations of genes (or gene complexes) such as those manifested in locally adapted landraces or breeds. The term genetic erosion is sometimes used in a narrow sense, such as for the loss of alleles or genes, as well as more broadly, referring to the loss of varieties or even species. The major driving forces behind genetic erosion in crops are: variety replacement, land clearing, overexploitation of species, population pressure, environmental degradation, overgrazing, policy and changing agricultural systems.

The main factor, however, is the replacement of local varieties by high yielding or exotic varieties or species. A large number of varieties can also often be dramatically reduced when commercial varieties (including GMOs) are introduced into traditional farming systems. Many researchers believe that the main problem related to agro-ecosystem management is the general tendency towards genetic and ecological uniformity imposed by the development of modern agriculture. Pressures for that ecological uniformity on farmers and breeders is caused by the food industry demand for more and more raw materials consistency.

In the case of Animal Genetic Resources for Food and Agriculture, major causes of genetic erosion are reported to include indiscriminate cross-breeding, increased use of exotic breeds, weak policies and institutions in animal genetic resources management, neglect of certain breeds because of a lack of profitability or competitiveness, the intensification of production systems, the effects of diseases and disease management, loss of pastures or other elements of the production environment, and poor control of inbreeding.

OVEREXPLOITATION

Overexploitation, also called overharvesting, refers to harvesting a renewable resource to the point of diminishing returns. Sustained overexploitation can lead to the destruction of the resource. The term applies to natural resources such as: wild medicinal plants, grazing pastures, game animals, fish stocks, forests, and water aquifers.

In ecology, overexploitation describes one of the five main activities threatening global biodiversity. Ecologists use the term to describe populations that are harvested at a rate that is unsustainable, given their natural rates of mortality and capacities for reproduction. This can result to extinction at the population level and even extinction of whole species. In conservation biology the term is usually used in the context of human economic activity that involves the taking of biological resources, or organisms, in larger numbers than their populations can withstand. The term is also used and defined somewhat differently in fisheries, hydrology, and natural resource management.

Overexploitation can lead to resource destruction, including extinctions. However it is also possible for overexploitation to be sustainable, as discussed below in the section on fisheries. In the context of fishing, the term overfishing can be used instead of overexploitation, as can overgrazing in stock management, overlogging in forest management, overdrafting in aquifer management, and endangered species in species monitoring. Overexploitation is not an activity limited to humans. Introduced predators and herbivores, for example, can overexploit native flora and fauna.

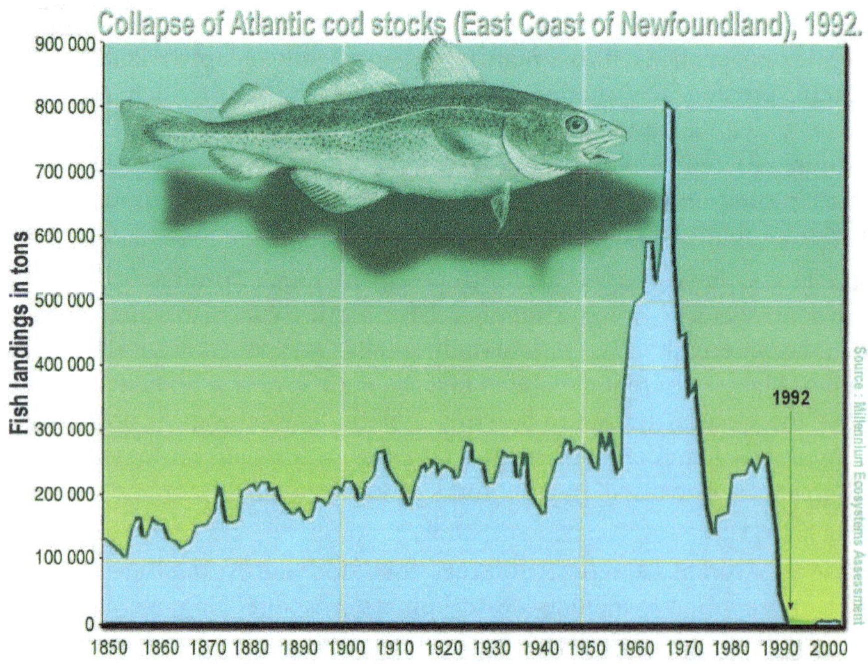

Atlantic cod stocks were severely overexploited in the 1970s and 1980s, leading to their abrupt collapse in 1992.

Concern about overexploitation is relatively recent, though overexploitation itself is not a new phenomenon. It has been observed for millennia. For example, ceremonial cloaks worn by the Hawaiian kings were made from the mamo bird; a single cloak used the feathers of 70,000 birds of this now-extinct species. The dodo, a flightless bird from

Mauritius, is another well-known example of overexploitation. As with many island species, it was naive about certain predators, allowing humans to approach and kill it with ease.

From the earliest of times, hunting has been an important human activity as a means of survival. There is a whole history of overexploitation in the form of overhunting. The overkill hypothesis (Quaternary extinction events) explains why the megafaunal extinctions occurred within a relatively short period of time. This can be traced with human migration. The most convincing evidence of this theory is that 80% of the North American large mammal species disappeared within 1000 years of the arrival of humans on the western hemisphere continents. The fastest ever recorded extinction of megafauna occurred in New Zealand, where by 1500 AD, just 200 years after settling the islands, ten species of the giant moa birds were hunted to extinction by the Māori. A second wave of extinctions occurred later with European settlement.

In more recent times, overexploitation has resulted in the gradual emergence of the concepts of sustainability and sustainable development, which has built on other concepts, such as sustainable yield, eco-development, and deep ecology.

Overexploitation doesn't necessarily lead to the destruction of the resource, nor is it necessarily unsustainable. However, depleting the numbers or amount of the resource can change its quality. For example, footstool palm is a wild palm tree found in Southeast Asia. Its leaves are used for thatching and food wrapping, and overharvesting has resulted in its leaf size becoming smaller.

Tragedy of the Commons

Cows on Selsley Common. The tragedy of the commons is a useful parable for understanding how overexploitation can occur.

The tragedy of the commons refers to a dilemma described in an article by that name written by Garrett Hardin and first published in the journal *Science* in 1968.

Central to Hardin's essay is an example which is a useful parable for understanding how overexploitation can occur. This example was first sketched in an 1833 pamphlet by William Forster Lloyd, as a hypothetical and simplified situation based on medieval land tenure in Europe, of herders sharing a common on which they are each entitled to let their cows graze. In Hardin's example, it is in each herder's interest to put each succeeding cow he acquires onto the land, even if the carrying capacity of the common is exceeded and it is temporarily or permanently damaged for all as a result. The herder receives all of the benefits from an additional cow, while the damage to the common is shared by the entire group. If all herders make this individually rational economic decision, the common will be overexploited or even destroyed to the detriment of all. However, since all herders reach the same rational conclusion, overexploitation in the form of overgrazing occurs, with immediate losses, and the pasture may be degraded to the point where it gives very little return.

Therein is the tragedy. Each man is locked into a system that compels him to increase his herd without limit—in a world that is limited. Ruin is the destination toward which all men rush, each pursuing his own interest in a society that believes in the freedom of the commons. (Hardin, 1968)

In the course of his essay, Hardin develops the theme, drawing in many examples of latter day commons, such as national parks, the atmosphere, oceans, rivers and fish stocks. The example of fish stocks had led some to call this the tragedy of the fishers. A major theme running through the essay is the growth of human populations, with the Earth's finite resources being the general common.

The tragedy of the commons has intellectual roots tracing back to Aristotle, who noted that what is common to the greatest number has the least care bestowed upon it, as well as to Hobbes and his *Leviathan*. The opposite situation to a tragedy of the commons is sometimes referred to as a tragedy of the anticommons: a situation in which rational individuals, acting separately, collectively waste a given resource by underutilizing it.

The tragedy of the commons can be avoided if it is appropriately regulated. Hardin's use of "commons" has frequently been misunderstood, leading Hardin to later remark that he should have titled his work "The tragedy of the unregulated commons".

Fisheries

In wild fisheries, overexploitation or overfishing occurs when a fish stock has been fished down "below the size that, on average, would support the long-term maximum sustainable yield of the fishery". However, overexploitation can be sustainable.

When a fishery starts harvesting fish from a previously unexploited stock, the biomass of the fish stock will decrease, since harvesting means fish are being removed. For sustainability, the rate at which the fish replenish biomass through reproduction must balance the rate at which the fish are being harvested. If the harvest rate is increased, then the stock biomass will further decrease. At a certain point, the maximum harvest yield that can be sustained will be reached, and further attempts to increase the harvest rate will result in the collapse of the fishery. This point is called the maximum sustainable yield, and in practice, usually occurs when the fishery has been fished down to about 30% of the biomass it had before harvesting started.

The Atlantic bluefin tuna is seriously overexploited. Scientists say 7,500 tons annually is the sustainable limit, yet the fishing industry continue to harvest 60,000 tons.

It is possible to fish the stock down further to, say, 15% of the pre-harvest biomass, and then adjust the harvest rate so the biomass remains at that level. In this case, the fishery is sustainable, but is now overexploited, because the stock has been run down to the point where the sustainable yield is less than it could be.

Fish stocks are said to "collapse" if their biomass declines by more than 95 percent of their maximum historical biomass. Atlantic cod stocks were severely overexploited in the 1970s and 1980s, leading to their abrupt collapse in 1992. Even though fishing has ceased, the cod stocks have failed to recover. The absence of cod as the apex predator in many areas has led to trophic cascades.

About 25% of world fisheries are now overexploited to the point where their current biomass is less than the level that maximizes their sustainable yield. These depleted fisheries can often recover if fishing pressure is reduced until the stock biomass returns to the optimal biomass. At this point, harvesting can be resumed near the maximum sustainable yield.

The tragedy of the commons can be avoided within the context of fisheries if fishing effort and practices are regulated appropriately by fisheries management. One effective approach may be assigning some measure of ownership in the form of individual transferable quotas (ITQs) to fishermen. In 2008, a large scale study of fisheries that used ITQs, and ones that didn't, provided strong evidence that ITQs help prevent collapses and restore fisheries that appear to be in decline.

Water Resources

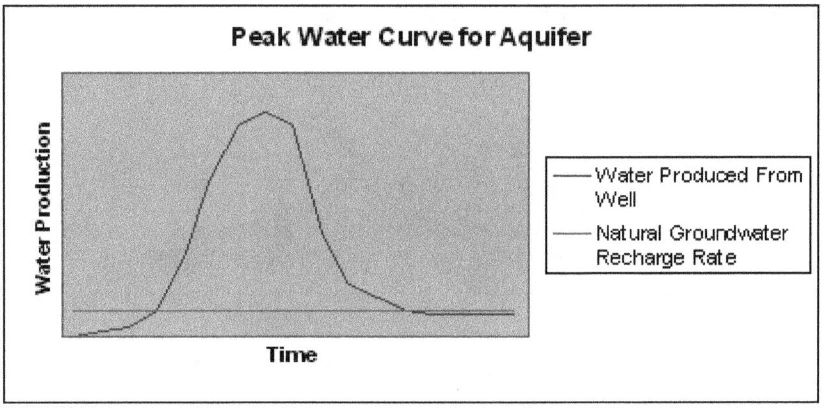

Overexploitation of groundwater from an aquifer can result to a peak water curve.

Water resources, such as lakes and aquifers, are usually renewable resources which naturally recharge (the term fossil water is sometimes used to describe aquifers which don't recharge). Overexploitation occurs if a water resource, such as the Ogallala Aquifer, is mined or extracted at a rate that exceeds the recharge rate, that is, at a rate that exceeds the practical sustained yield. Recharge usually comes from area streams, rivers and lakes. An aquifer which has been overexploited is said to be overdrafted or depleted. Forests enhance the recharge of aquifers in some locales, although generally forests are a major source of aquifer depletion. Depleted aquifers can become polluted with contaminants such as nitrates, or permanently damaged through subsidence or through saline intrusion from the ocean.

This turns much of the world's underground water and lakes into finite resources with peak usage debates similar to oil. These debates usually centre around agriculture and suburban water usage but generation of electricity from nuclear energy or coal and tar sands mining is also water resource intensive. A modified Hubbert curve applies to any resource that can be harvested faster than it can be replaced. Though Hubbert's original analysis did not apply to renewable resources, their overexploitation can result in a Hubbert-like peak. This has led to the concept of peak water.

Forest Resources

Forests are overexploited when they are logged at a rate faster than reforestation takes place. Reforestation competes with other land uses such as food production, livestock grazing, and living space for further economic growth. Historically utilization of forest products, including timber and fuel wood, have played a key role in human societies, comparable to the roles of water and cultivable land. Today, developed countries continue to utilize timber for building houses, and wood pulp for paper. In developing countries, almost three billion people rely on wood for heating and cooking. Short-term economic gains made by conversion of forest to agriculture, or overexploitation of wood products, typically leads to loss of long-term income and long term biological productivity. West Africa, Madagascar, Southeast Asia and many other regions have experienced lower revenue because of overexploitation and the consequent declining timber harvests.

Beech forest – Grib Skov, Denmark

Biodiversity

The rich diversity of marine life inhabiting coral reefs attracts bioprospectors. Many coral reefs are overexploited; threats include coral mining, cyanide and blast fishing, and overfishing in general.

Overexploitation is one of the main threats to global biodiversity. Other threats include pollution, introduced and invasive species, habitat fragmentation, habitat destruction, uncontrolled hybridization, global warming, ocean acidification, and the driver behind many of these, human overpopulation.

One of the key health issues associated with biodiversity is drug discovery and the availability of medicinal resources. A significant proportion of drugs are natural products derived, directly or indirectly, from biological sources. Marine ecosystems are of particular interest in this regard. However, unregulated and inappropriate bioprospecting could potentially lead to overexploitation, ecosystem degradation, and loss of biodiversity.

Endangered Species

It is not just humans that overexploit their resources. Overgrazing can occur naturally, caused by native fauna, as shown in the upper right.

Overexploitation threatens one-third of endangered vertebrates, as well as other groups. Excluding edible fish, the illegal trade in wildlife is valued at $10 billion per year. Industries responsible for this include the trade in bushmeat, the trade in Chinese medicine, and the fur trade. The Convention for International Trade in Endangered Species of Wild Fauna and Flora, or CITES was set up in order to control and regulate the trade in endangered animals. It currently protects, to a varying degree, some 33,000 species of animals and plants. It is estimated that a quarter of the endangered vertebrates in the United States of America and half of the endangered mammals is attributed to overexploitation.

All living organisms require resources to survive. Overexploitation of these resources for protracted periods can deplete natural stocks to the point where they are unable to recover within a short time frame. Humans have always harvested food and other resources they have needed to survive. Human populations, historically, were small, and methods of collection limited to small quantities. With an exponential increase in human population, expanding markets and increasing demand, combined with improved access and techniques for capture, are causing the exploitation of many species beyond sustainable levels. In practical terms, if continued, it reduces valuable resources to such low levels that their exploitation is no longer sustainable and can lead to the extinction of a species, in addition to having dramatic, unforeseen effects, on the ecosystem. Overexploitation often occurs rapidly as markets open, utilising previously untapped resources, or locally used species.

The Carolina parakeet was hunted to extinction.

Today, overexploitation and misuse of natural resources is an ever-present threat for species richness. This is more prevalent when looking at island ecology and the species that inhabit them, as islands can be viewed as the world in miniature. Island endemic populations are more prone to extinction from overexploitation, as they often exist at low densities with reduced reproductive rates. A good example of this are island snails, such as the Hawaiian *Achatinella* and the French Polynesian *Partula*. Achatinelline snails have 15 species listed as extinct and 24 critically endangered while 60 species of partulidae are considered extinct with 14 listed as critically endangered. The WCMC have attributed over-collecting and very low lifetime fecundity for the extreme vulnerability exhibited among these species.

As another example, when the humble hedgehog was introduced to the Scottish island of Uist, the population greatly expanded and took to consuming and overexploiting shorebird eggs, with drastic consequences for their breeding success. Twelve species of avifauna are affected, with some species numbers being reduced by 39%.

Where there is substantial human migration, civil unrest, or war, controls may no longer exist. With civil unrest, for example in the Congo and Rwanda, firearms have become common and the breakdown of food distribution networks in such countries leaves the resources of the natural environment vulnerable. Animals are even killed as target practice, or simply to spite the government. Populations of large primates, such as gorillas and chimpanzees, ungulates and other mammals, may be reduced by 80% or more by hunting, and certain species may be eliminated altogether. This decline has been called the bushmeat crisis.

Overall, 50 bird species that have become extinct since 1500 (approximately 40% of the total) have been subject to overexploitation, including:

- Great Auk – The penguin-like bird of the north, hunted for its feathers, meat, fat and oil.
- Carolina parakeet – The only parrot species native to the eastern United States, was hunted for crop protection and its feathers.

Other species affected by overexploitation include:

- The international trade in fur: chinchilla, vicuña, giant otter, and numerous cat species.
- Insect collectors: butterflies
- Horticulturists: New Zealand mistletoe (*Trilepidia adamsii*), orchids, cacti, and many other plant species.
- Shell collectors: Marine molluscs
- Aquarium hobbyists: tropical fish
- Chinese medicine: bears, tigers, rhinos, seahorses, Asian black bear, and saiga antelope
- Novelty pets: snakes, parrots, primates, and big cats

Cascade Effects

Overexploitation of species can result to knock-on or cascade effects. This can particularly apply if, through overexploitation, a habitat loses its apex predator. Because of the loss of the top predator, a dramatic increase in their prey species can occur. In turn, the unchecked prey can then overexploit their own food resources until population numbers dwindle, possibly to the point of extinction.

A classic example of cascade effects occurred with sea otters. Starting before the 17th century and not phased out until 1911, sea otters were hunted aggressively for their exceptionally warm and valuable pelts, which could fetch up to $2500 US. This caused cascade effects through the kelp forest ecosystems along the Pacific Coast of North America.

Overexploiting sea otters resulted in cascade effects which destroyed kelp forest ecosystems.

One of the sea otters' primary food sources is the sea urchin. When hunters caused sea otter populations to decline, an ecological release of sea urchin populations occurred. The sea urchins then overexploited their main food source, kelp, creating urchin barrens, areas of seabed denuded of kelp, but carpeted with urchins. No longer having food to eat, the sea urchin became locally extinct as well. Also, since kelp forest ecosystems are homes to many other species, the loss of the kelp caused other cascade effects of secondary extinctions.

In 1911, when only one small group of 32 sea otters survived in a remote cove, an international treaty was signed to prevent further exploitation of the sea otters. Under heavy protection, the otters multiplied and repopulated the depleted areas, which slowly recovered. More recently, with declining numbers of fish stocks, again due to overexploitation, killer whales have experienced a food shortage and have been observed feeding on sea otters, again reducing their numbers.

OVERGRAZING

Overgrazing occurs when plants are exposed to intensive grazing for extended periods of time, or without sufficient recovery periods. It can be caused by either livestock in poorly managed agricultural applications, game reserves, or nature reserves. It can also be caused by immobile, travel restricted populations of native or non-native wild animals. The only thing to stop overgrazing is to limit the animals space to roam around. Cows are the main things that causes overgrazing and overgrazing can lead to desertification.

It reduces the usefulness, productivity, and biodiversity of the land and is one cause of desertification and erosion. Overgrazing is also seen as a cause of the spread of invasive species of non-native plants and of weeds. It is not caused by nomadic grazers in huge populations of travel herds, such as the American bison of the Great Plains, or migratory Wildebeests of the African savannas, nor by holistic planned grazing.

Satellite image of the border between Israel and Egypt. The Egyptian side, to the left, is overgrazed.

Preventing Overgrazing

Preventing Man-Made Overgrazing

Immobilized goat herd inside of a pen in an overgrazed habitation of Norte Chico, Chile.

Huge herd of migratory Wildebeest in Masai Mara during the Great Migration.

Sustainable grassland production is based on grass and grassland management, land management, animal management, and livestock marketing. Grazing management, with sustainable agriculture and agroecology practices, is the foundation of grassland-based livestock production since it affects both animal and plant health and productivity. There are several new grazing models and management systems that attempt to reduce or eliminate overgrazing like Holistic management and Permaculture

Indicators

One indicator of overgrazing is that the animals run short of pasture. In some regions of the United States under continuous grazing, overgrazed pastures are predominated by short-grass species such as bluegrass and will be less than 2-3 inches tall in the grazed areas. In other parts of the world, overgrazed pasture is typically taller than sustainably grazed pasture, with grass heights typically over 1 meter and dominated by unpalatable species such as *Aristida* or *Imperata*. In all cases, palatable tall grasses such as orchard grass are sparse or non-existent. In such cases of overgrazing, soil may be visible between plants in the stand, allowing erosion to occur, though in many circumstances overgrazed pastures have a greater sward cover than sustainably grazed pastures.

Rotational Grazing

Under rotational grazing, overgrazed plants do not have enough time to recover to the proper height between grazing events. The animals resume grazing before the plants have restored carbohydrate reserves and grown back roots lost after the last defoliation. The result is the same as under continuous grazing: in some parts of the United States tall-growing species die and short-growing species that are more subject to drought injury predominate the pasture, while in most other parts of the world tall, drought tolerant, unpalatable species such as *Imperata* or *Aristida* come to dominate. As the sod thins, weeds encroach into the pasture in some parts of the United States, whereas in most other parts of the worlds overgrazing can promote thick swards of native unpalatable grasses that hamper the spread of weeds.

Another indicator of overgrazing in some parts of North America is that livestock run out of pasture, and hay needs to be fed early in the fall. In contrast, most areas of the world do not experience the same climatic regime as the continental United States and hay feeding is rarely conducted.

Overgrazing is also indicated in livestock performance and condition. Cows having inadequate pasture immediately following their calf's weaning may have poor body condition the following season. This may reduce the health and vigor of cows and calves at calving. Also, cows in poor body condition do not cycle as soon after calving, which can result to delayed breeding. This can result to a long calving season. With good cow genetics, nutrition, ideal seasons and controlled breeding 55% to 75% of the calves should come in the first 21 days of the calving season. Poor weaning weights of calves can be caused by insufficient pasture, when cows give less milk and the calves need pasture to maintain weight gain.

Ecological Impacts

Overgrazing typically increases soil erosion. Reduction in soil depth, soil organic matter and soil fertility impair the land's future natural and agricultural productivity. Soil fertility can sometimes be mitigated by applying the appropriate lime and organic fertilizers. However, the loss of soil depth and organic matter takes centuries to correct. Their loss is critical in determining the soil's water-holding capacity and how well pasture plants do during dry weather.

Native plant grass species, both individual bunch grasses and in grasslands, are especially vulnerable.

North America - United States

In the continental United States, to prevent overgrazing, match the forage supplement to the herd's requirement. This means that a buffer needs to be in the system to adjust for the fastest growth of forages.

Another potential buffer is to plant warm-season perennial grasses such as switchgrass, which do not grow early in the season. This reduces the area that the livestock can use early in the season, making it easier for them to keep up with the cool-season grasses. The animals then use the warm-season grasses during the heat of the summer, and the cool-season grasses recover for fall grazing.

The grazing guidelines in the table are for rotationally grazed, cool-season forages. When using continuous grazing, manage pasture height at one-half the recommended turn-in height for rotational grazing to optimize plant health. The growth habit of some forage species, such as alfalfa, does not permit their survival under continuous grazing. When managing for legumes in the stand, it is beneficial to use rotational grazing and graze the stand close and then give adequate rest to stimulate the legumes' growth.

Economic Theory

Overgrazing is used as an example in the economic concept now known as the Tragedy of the Commons devised in a 1968 paper by Garrett Hardin. This cited the work of a Victorian economist who used the over-grazing of common land as an example of behaviour. Hardin's example could only apply to unregulated use of land regarded as a common resource.

Normally, rights of use of Common land in England and Wales were, and still are, closely regulated, and available only to "commoners". If excessive use was made of common land, for example in overgrazing, a common would be "stinted", that is, a limit would be put on the number of animals each commoner was allowed to graze. These regulations were responsive to demographic and economic pressure; thus rather than let a common become degraded, access was restricted even further. This important part of actual historic practice was absent from the economic model of Hardin. In reality the use of common land in England and Wales was a triumph of conserving a scarce resource using agreed custom and practice.

Africa-Sahel Region

- overgrazing consequences in region: 2010 Sahel famine

Australia

Overgrazing does not necessarily mean total denudation of the land. Overgrazing can also occur with native species. In the Australian Capital Territory, Australia the local government, in 2013, authorised the culling of 1455 kangaroos due to overgrazing.

REFERENCES

1. Frison, E.A.; Cherfas, J.; Hodgkin, T. Agricultural Biodiversity Is Essential for a Sustainable Improvement in Food and Nutrition Security. Sustainability 2011, 3, 238-253.
2. FAO, (1996). Global Plan of Action for the Conservation and Sustainable Utilization of Plant Genetic Resources for Food and Agriculture. Food and Agriculture Organization of the United Nations, "Archived copy". Archived from the original on 2008-09-20. Retrieved 2008-09-05.
3. FAO : SD Dimensions : Environment : Environmental conventions and agreements Archived December 13, 2004, at the Wayback Machine.
4. Jackson, L., Bawa, K., Pascual, U., and Perrings, C. (2005).agroBIODIVERSITY: A new science agenda for biodiversity in support of sustainable agroecosystems. DIVERSITAS Report N°4. 40 pp. "Archived copy" (PDF). Archived from the original (PDF) on 2011-07-24. Retrieved 2010-02-27.

5. FAO, (1997). The State of the World's Plant Genetic Resources for Food and Agriculture. Food and Agriculture Organization of the United Nations, "Archived copy" (PDF). Archived from the original (PDF) on 2006-10-27. Retrieved 2008-09-05.
6. Gao, L.Z., (2003). The conservation of rice biodiversity in China: significance, genetic erosion, ethnobotany and prospect. Genetic Resources and Crop Evolution, 50: 17-32.
7. Guarino, L., (1995). Assessing the threat of genetic erosion. In: Collecting Plant Genetic Diversity: Technical Guidelines (eds. Guarino L, Rao VR, Reid R) pp. 67-74. CAB International, Wallingford.
8. FAO. 2015. The Second Report on the State of the World's Animal Genetic Resources for Food and Agriculture. Rome.
9. Virchow, Detlef. Conservation of genetic resources: Costs and implications for a sustainable utilization of plant genetic resources for food and agriculture Springer, 1999. p22
10. Eric Elsner. "Genetic Resources and Genetic Diversity". Retrieved 29 October 2014.
11. Kloppenburg, Jack Ralph Jr. "First the Seed: The political economy of plant biotechnology, 2nd edition" University of Wisconsin Press 2004. 163
12. Khoury, C.K.; Bjorkman, A.D.; Dempewolf, H.; Ramirez-Villegas, J.; Guarino, L.; Jarvis, A.; Rieseberg, L.H.; Struik, P.C. (2014). "Increasing homogeneity in global food supplies and the implications for food security". PNAS. 111 (11): 4001–4006. doi:10.1073/pnas.1313490111.
13. Kinver, Mark. "Crop diversity decline 'threatens food security'". BBC. Retrieved 13 June 2016.
14. Fischetti, Mark. "Diets around the world are becoming more similar". Scientific American. p. 72. Retrieved 13 June 2016.
15. Thrupp, L. A. (2000) 'Linking agricultural biodiversity and food security: the valuable role of agrobiodiversity for sustainable agriculture', International Affairs, 76(2): 265-281.pdf
16. Galluzi, G., Eyzaguirre, P. and Negri, V. (2010) 'Home gardens: neglected hotspots of agro-biodiversity and cultural diversity', Biodiversity and Conservation, 19: 3635-3654.
17. Kull, C. A., Carriere, S. M., Moreau, S., Rakoto Ramiarantsoa, H., Blanc-Pamard, C. and Tassin, J. (2013) 'Melting pots of biodiversity: tropical smallholder farm landscapes as guarantors of sustainability', Environment, 55(2): 6-15.
18. Bengtsoon, J., et al. (2005). The effects of organic agriculture on biodiversity and abundance: a metaanalysis. Journal of Applied Ecology, 42: 261–269.
19. Morandin, Lora A., and Mark L. Winston. (2005). WILD BEE ABUNDANCE AND SEED PRODUCTION IN CONVENTIONAL, ORGANIC, AND GENETICALLY MODIFIED CANOLA. Ecological Applications, 15:871–881.
20. Kenneth T. Frank; Brian Petrie; Jae S. Choi; William C. Leggett (2005). "Trophic

Cascades in a Formerly Cod-Dominated Ecosystem". Science. 308 (5728): 1621–1623. doi:10.1126/science.1113075. PMID 15947186.

21. Wilcove, D. S.; Rothstein, D.; Dubow, J.; Phillips, A.; Losos, E. (1998). "Quantifying threats to imperiled species" (PDF). BioScience. 48: 607–615. doi:10.2307/1313420.
22. Oxford. (1996). Oxford Dictionary of Biology. Oxford University Press.
23. Holdaway, R. N.; Jacomb, C. (2000). "Rapid Extinction of the Moas (Aves: Dinornithiformes): Model, Test, and Implications" (PDF). Science. 287 (5461): 2250–2254. doi:10.1126/science.287.5461.2250. PMID 10731144.
24. Tennyson, A.; Martinson, P. (2006). Extinct Birds of New Zealand. Wellington, New Zealand: Te Papa Press. ISBN 978-0-909010-21-8.
25. Fryer, Jonathan (2002-09-14). "Bringing the dodo back to life". BBC News. Retrieved 2006-09-07.
26. Paul S. Martin
27. Larkin, P. A. (1977). "An epitaph for the concept of maximum sustained yield". Transactions of the American Fisheries Society. 106 (1): 1–11. doi:10.1577/1548-8659(1977)106<1:AEFTCO>2.0.CO;2.
28. Lubchenco, J. (1991). "The Sustainable Biosphere Initiative: An ecological research agenda". Ecology. 72: 371–412. doi:10.2307/2937183.
29. Lee, K. N. (2001). "Sustainability, concept and practice of". In Levin, S. A. Encyclopedia of Biodiversity. 5. San Diego, CA: Academic Press. pp. 553–568. ISBN 0-12-226864-4.
30. Naess, A. (1986). "Intrinsic value: Will the defenders of nature please rise?". In Soulé, M. E. Conservation Biology: The Science of Scarcity and Diversity. Sunderland, MA: Sinauer Associates. pp. 153–181. ISBN 0-87893-794-3.
31. Sessions, G., ed. (1995). Deep Ecology for the 21st Century: Readings on the Philosophy and Practice of the New Environmentalism. Boston: Shambala Books. ISBN 1-57062-049-0.
32. Hardin, Garrett (1968). "The Tragedy of the Commons". Science. 162 (3859): 1243–1248. doi:10.1126/science.162.3859.1243. PMID 5699198. Also available here.
33. Bowles, Samuel (2004). Microeconomics: Behavior, Institutions, and Evolution. Princeton University Press. pp. 27–29. ISBN 0-691-09163-3.
34. Ostrom, E. (1992). "The rudiments of a theory of the origins, survival, and performance of common-property institutions". In Bromley, D. W. Making the Commons Work: Theory, Practice and Policy. San Francisco: ICS Press.
35. Feeny, D.; et al. (1990). "The Tragedy of the Commons: Twenty-two years later". Human Ecology. 18 (1): 1–19. doi:10.1007/BF00889070.
36. "Will commons sense dawn again in time?". The Japan Times Online.
37. NOAA: FishWatch glossary Retrieved 2 February 2010.

38. Bolden, E.G., Robinson, W.L. (1999), Wildlife ecology and management 4th ed. Prentice-Hall, Inc. Upper Saddle River, NJ. ISBN 0-13-840422-4
39. Grafton, RQ; Kompas, T; Hilborn, RW (2007). "Economics of Overexploitation Revisited". Science. 318 (5856): 1601. Bibcode:2007Sci...318.1601G. doi:10.1126/science.1146017.
40. Rosenberg, AA (2003). "Managing to the margins: the overexploitation of fisheries". Frontiers in Ecology and the Environment. 1 (2): 102–106. doi:10.1890/1540-9295(2003)001[0102:MTTMTO]2.0.CO;2.
41. New Scientist: Guaranteed fish quotas halt commercial free-for-all
42. A Rising Tide: Scientists find proof that privatising fishing stocks can avert a disaster The Economist, 18th Sept, 2008.
43. Meena Palaniappan & Peter H. Gleick (2008). "The World's Water 2008-2009, Ch 1." (PDF). Pacific Institute. Retrieved 2009-01-31.
44. "Underlying Causes of Deforestation: UN Report". Archived from the original on 2001-04-11.
45. http://www.azstarnet.com/sn/byauthor/244797
46. http://www.uswaternews.com/archives/arcsupply/6worllarg2.html
47. http://www.earth-policy.org/Updates/2005/Update47_data.htm
48. http://www.epa.gov/cleanrgy/water_resource.htm
49. http://atlas.aaas.org/pdf/63-66.pdf Forest Products
50. "Destruction of Renewable Resources".
51. Rhymer, Judith M.; Simberloff, Daniel (1996). "Extinction by Hybridization and Introgression". Annual Review of Ecology and Systematics. 27: 83–109. doi:10.1146/annurev.ecolsys.27.1.83. JSTOR 2097230.
52. Kannan, R.; James, D. A. (2009). "Effects of climate change on global biodiversity: a review of key literature" (PDF). Tropical Ecology. 50 (1): 31–39. ISSN 0564-3295. Retrieved 2014-05-21.
53. Mora, C.; et al. (2013). "Biotic and Human Vulnerability to Projected Changes in Ocean Biogeochemistry over the 21st Century". PLoS Biology. 11: e1001682. doi:10.1371/journal.pbio.1001682. PMC 3797030 . PMID 24143135.
54. Dumont, E. (2012). "Estimated impact of global population growth on future wilderness extent." (PDF). Earth System Dynamics Discussions. 3: 433–452. Bibcode:2012ESDD....3..433D. doi:10.5194/esdd-3-433-2012.
55. (2006) "Molecular Pharming" GMO Compass Retrieved November 5, 2009, From http://www.gmocompass.org/eng/search/
56. Roopesh, J.; et al. (2008). "Marine organisms: Potential Source for Drug Discovery" (PDF). Current Science. 94 (3): 292.
57. Dhillion, S. S.; Svarstad, H.; Amundsen, C.; Bugge, H. C. (September 2002). "Bioprospecting: Effects on Environment and Development". AMBIO. 31 (6): 491–493. doi:10.1639/0044-7447(2002)031[0491:beoead]2.0.co;2. JSTOR 4315292. PMID 12436849.

58. Cole, Andrew (2005). "Looking for new compounds in sea is endangering ecosystem". BMJ. 330 (7504): 1350. doi:10.1136/bmj.330.7504.1350-d.
59. "COHAB Initiative - on Natural Products and Medicinal Resources". Cohabnet.org. Retrieved 2009-06-21.
60. Hemley 1994.
61. Primack, R. B. (2002). Essentials of Conservation Biology (3rd ed.). Sunderland: Sinauer Associates. ISBN 0-87893-719-6.
62. Redford 1992, Fitzgibon et al. 1995, Cuarón 2001.
63. Frankham, R.; Ballou, J. D.; Briscoe, D. A. (2002). Introduction to Conservation Genetics. New York: Cambridge University Press. ISBN 0-521-63014-2.
64. Dowding, J. E.; Murphy, E. C. (2001). "The Impact of Predation be Introduced Mammals on Endemic Shorebirds in New Zealand: A Conservation Perspective". Biological Conservation. 99 (1): 47–64. doi:10.1016/S0006-3207(00)00187-7.
65. "IUCN Red List". 2003b.
66. "IUCN Red List". 2003c. Retrieved 9 December 2003.
67. WCMC. (1992). McComb, J., Groombridge, B., Byford, E., Allan, C., Howland, J., Magin, C., Smith, H., Greenwood, V. and Simpson, L. (1992). World Conservation Monitoring Centre. Chapman and Hall.
68. Jackson, D. B.; Fuller, R. J.; Campbell, S. T. (2004). "Long-term Population Changes Among Breeding Shorebirds in the Outer Hebrides, Scotland, In Relation to Introduced Hedgehogs (Erinaceus europaeus)". Biological Conservation. 117 (2): 151–166. doi:10.1016/S0006-3207(03)00289-1.
69. Jones, R. F. (1990). "Farewell to Africa". Audubon. 92: 1547–1551.
70. Wilkie, D. S.; Carpenter, J. F. (1999). "Bushmeat hunting in the Congo Basin: An assessment of impacts and options for migration". Biodiversity and Conservation. 8 (7): 927–955. doi:10.1023/A:1008877309871.
71. The LUCN Red List of Threatened Species (2009).
72. Collins, Nick. "Chinese medicines contain traces of endangered animals". Telegraph.
73. Estes, J. A.; Duggins, D. O.; Rathbun, G. B. (1989). "The ecology of extinctions in kelp forest communities". Conservation Biology. 3 (3): 251–264. doi:10.1111/j.1523-1739.1989.tb00085.x.
74. Dayton, P. K.; Tegner, M. J.; Edwards, P. B.; Riser, K. L. (1998). "Sliding baselines, ghosts, and reduced expectations in kelp forest communities". Ecol. Appl. 8 (2): 309–322. doi:10.1890/1051-0761(1998)008[0309:SBGARE]2.0.CO;2.
75. Krebs, C. J. (2001). Ecology (5th ed.). San Francisco: Benjamin Cummings. ISBN 0-321-04289-1.
76. Laduke, Winona (1999). All Our Relations: Native Struggles for Land and Life (PDF). Cambridge, MA: South End Press. p. 146. ISBN 0896085996. Retrieved 30 March 2015.

77. Duval, Clay. "Bison Conservation: Saving an Ecologically and Culturally Keystone Species" (PDF). Duke University. Retrieved April 13, 2015.
78. In balance with, and accompanied by, prides of keystone predators.
79. "Holistic Land Management: Key to Global Stability" by Terry Waghorn. Forbes. 20 December 2012.
80. Savory, Allan. "How to green the world's deserts and reverse climate change". youtube.com. TED. Retrieved 14 September 2015.
81. Savory, Allan. "Can sheep save the planet?". youtube.com. IWTOCHANNEL. Retrieved 14 September 2015.
82. West Virginia University Extension Service Overgrazing Can Hurt Environment, Your Pocketbook Ed Rayburn. 2000.
83. C.Michael Hogan. 2009. Overgrazing. Encyclopedia of Earth. Sidney Draggan, topic ed.; Cutler J. Cleveland, ed., National council for Science and the Environment, Washington DC
84. Garrett Hardin, "The Tragedy of the Commons", Science, Vol. 162, No. 3859 (December 13, 1968), pp. 1243-1248. Also available here and here.
85. Susan Jane Buck Cox - "No tragedy on the Commons" Journal of Environmental Ethics, Vol 7, Spring 1985

CHAPTER 4

Ecological Effects of Biodiversity

Chapter Contents

Biodiversity is the foundation of ecosystem services to which human well-being is intimately linked. No feature of Earth is more complex, dynamic, and varied than the layer of living organisms that occupy its surfaces and its seas, and no feature is experiencing more dramatic change at the hands of humans than this extraordinary, singularly unique feature of Earth.

INTRODUCTION

The diversity of species and genes in ecological communities affects the functioning of these communities. These ecological effects of biodiversity in turn are affected by both climate change through enhanced greenhouse gases, aerosols, and loss of land cover, and biological diversity, causing a rapid loss of ecosystems and extinctions of species and local populations. The current rate of extinction is sometimes considered a mass extinction, with current species extinction rates on the order of 100 to 1000 times as high as in the past.

The two main areas where the effect of biodiversity on ecosystem function have been studied are the relationship between diversity and productivity, and the relationship between diversity and community stability. More biologically diverse communities appear to be more productive (in terms of biomass production) than are less diverse communities, and they appear to be more stable in the face of perturbations.

Also, animals that inhabit an area may alter the surviving conditions by factors assimilated by climate.

Definitions

In order to understand the effects that changes in biodiversity will have on ecosystem functioning, it is important to define some terms. Biodiversity is not easily defined, but may be thought of as the number and/or evenness of genes, species, and ecosystems in a region. This definition includes genetic diversity, or the diversity of genes within a species, species diversity, or the diversity of species within a habitat or region, and ecosystem diversity, or the diversity of habitats within a region.

Two things commonly measured in relation to changes in diversity are productivity and stability. Productivity is a measure of ecosystem function. It is generally measured by taking the total aboveground biomass of all plants in an area. Many assume that it can be used as a general indicator of ecosystem function and that total resource use and other indicators of ecosystem function are correlated with productivity.

Stability is much more difficult to define, but can be generally thought of in two ways. General stability of a population is a measure that assumes stability is higher if there is less of a chance of extinction. This kind of stability is generally measured by measuring the variability of aggregate community properties, like total biomass, over time. The other definition of stability is a measure of resilience and resistance, where an ecosystem that returns quickly to an equilibrium after a perturbation or resists invasion is thought of as more stable than one that doesn't.

GENETIC DIVERSITY

Genetic diversity is the total number of genetic characteristics in the genetic makeup of a species. It is distinguished from genetic variability, which describes the tendency of genetic characteristics to vary.

Genetic diversity serves as a way for populations to adapt to changing environments. With more variation, it is more likely that some individuals in a population will possess variations of alleles that are suited for the environment. Those individuals are more likely to survive to produce offspring bearing that allele. The population will continue for more generations because of the success of these individuals.

The academic field of population genetics includes several hypotheses and theories regarding genetic diversity. The neutral theory of evolution proposes that diversity is the result of the accumulation of neutral substitutions. Diversifying selection is the hypothesis that two subpopulations of a species live in different environments that select for different alleles at a particular locus. This may occur, for instance, if a species has a large range relative to the mobility of individuals within it. Frequency-dependent selection is the hypothesis that as alleles become more common, they become more vulnerable. This occurs in host–pathogen interactions, where a high frequency of a defensive allele among the host means that it is more likely that a pathogen will spread if it is able to overcome that allele.

Importance of Genetic Diversity

A 2007 study conducted by the National Science Foundation found that genetic diversity and biodiversity (in terms of species diversity) are dependent upon each other—that diversity within a species is necessary to maintain diversity among species, and vice versa. According to the lead researcher in the study, Dr. Richard Lankau, "If any one type is removed from the system, the cycle can break down, and the community becomes dominated by a single species." Genotypic and phenotypic diversity have been found in all species at the protein, DNA, and organismal levels; in nature, this diversity is nonrandom, heavily structured, and correlated with environmental variation and stress.

The interdependence between genetic and species diversity is delicate. Changes in species diversity lead to changes in the environment, leading to adaptation of the remaining species. Changes in genetic diversity, such as in loss of species, leads to a loss of biological diversity. Loss of genetic diversity in domestic animal populations has also been studied and attributed to the extension of markets and economic globalization.

Survival and Adaptation

Genetic diversity plays an important role in the survival and adaptability of a species. When a population's habitat changes, the population may have to adapt to survive; the ability of the population to adapt to the changing environment will determine their ability to cope with an environmental challenge. Variation in the population's gene pool provides variable traits among the individuals of that population. These variable traits can be selected for, via natural selection, ultimately leading to an adaptive change in the population, allowing it to survive in the changed environment. If a population of a species has a very diverse gene pool then there will be more variety in the traits of individuals of that population and consequently more traits for natural selection to act upon to select the fittest individuals to survive.

Genetic diversity is essential for a species to evolve. With very little gene variation within the species, healthy reproduction becomes increasingly difficult, and offspring are more likely to have problems resulting from inbreeding. The vulnerability of a population to certain types of diseases can also increase with reduction in genetic diversity. Concerns about genetic diversity are especially important with large mammals due to their small population size and high levels of human-caused population effects.

Agricultural Relevance

When humans initially started farming, they used selective breeding to pass on desirable traits of the crops while omitting the undesirable ones. Selective breeding leads to monocultures: entire farms of nearly genetically identical plants. Little to no genetic diversity makes crops extremely susceptible to widespread disease; bacteria morph and change constantly and when a disease-causing bacterium changes to attack a specific genetic variation, it can easily wipe out vast quantities of the species. If the genetic variation that the bacterium is best at attacking happens to be that which humans have selectively bred to use for harvest, the entire crop will be wiped out.

A very similar occurrence is the cause of the infamous Potato Famine in Ireland. Since new potato plants do not come as a result of reproduction, but rather from pieces of the parent plant, no genetic diversity is developed, and the entire crop is essentially a clone of one potato, it is especially susceptible to an epidemic. In the 1840s, much of Ireland's population depended on potatoes for food. They planted namely the "lumper" variety of potato, which was susceptible to a rot-causing oomycete called *Phytophthora infestans*. This oomycete destroyed the vast majority of the potato crop, and left one million people to starve to death.

Livestock Biodiversity

The genetic diversity of livestock species allows animal production to be practiced in a range of environments and with a range of different objectives. It provides the raw material for selective breeding programmes and allows livestock populations to adapt as environmental conditions change.

Livestock biodiversity can be lost as a result of breed extinctions and other forms of genetic erosion. As of June 2014, among the 8774 breeds recorded in the Domestic Animal Diversity Information System (DAD-IS), operated by the Food and Agriculture Organization of the United Nations (FAO), 17 percent were classified as being at risk of extinction and 7 percent already extinct. There is now a Global Plan of Action for Animal Genetic Resources that was developed under the auspices of the Commission on Genetic Resources for Food and Agriculture in 2007, that provides a framework and guidelines for the management of animal genetic resources.

Awareness of the importance of maintaining animal genetic resources has increased over time. FAO has published two reports on the state of the world's animal genetic resources for food and agriculture, which cover detailed analyses of our global livestock diversity and ability to manage and conserve them.

Coping with Low Genetic Diversity

Photomontage of planktonic organisms.

A Tanzanian cheetah.

The natural world has several ways of preserving or increasing genetic diversity. Among oceanic plankton, viruses aid in the genetic shifting process. Ocean viruses, which infect the plankton, carry genes of other organisms in addition to their own. When a virus containing the genes of one cell infects another, the genetic makeup of the latter changes. This constant shift of genetic makeup helps to maintain a healthy population of plankton despite complex and unpredictable environmental changes.

Cheetahs are a threatened species. Low genetic diversity and resulting poor sperm quality has made breeding and survivorship difficult for cheetahs. Moreover, only about 5% of cheetahs survive to adulthood. However, it has been recently discovered that female cheetahs can mate with more than one male per litter of cubs. They undergo induced ovulation, which means that a new egg is produced every time a female mates. By mating with multiple males, the mother increases the genetic diversity within a single litter of cubs.

Measures of Genetic Diversity

Genetic Diversity of a population can be assessed by some simple measures.

- **Gene Diversity** is the proportion of polymorphic loci across the genome.
- **Heterozygosity** is the fraction of individuals in a population that are heterozygous for a particular locus.
- **Alleles per locus** is also used to demonstrate variability.
- **Nucleotide diversity** is the extent of nucleotide polymorphisms within a population, and is commonly measured through molecular markers such as micro- and minisatellite sequences, mitochondrial DNA, and single-nucleotide polymorphisms (SNPs).

Other Measures of Diversity

Alternatively, other types of diversity may be assessed for organisms:

- species diversity
- ecological diversity
- morphological diversity
- degeneracy

There are broad correlations between different types of diversity. For example, there is a close link between vertebrate taxonomic and ecological diversity.

SPECIES DIVERSITY

Species diversity is the number of different species that are represented in a given community (a dataset). The effective number of species refers to the number of equally abundant species needed to obtain the same mean proportional species abundance as that observed in the dataset of interest (where all species may not be equally abundant). Species diversity consists of two components: species richness and species evenness. Species richness is a simple count of species, whereas species evenness quantifies how equal the abundances of the species are.

Calculation of Diversity

Species diversity in a dataset can be calculated by first taking the weighted average of species proportional abundances in the dataset, and then taking the inverse of this. The equation is:

$$^{q}D = \frac{1}{\sqrt[q-1]{\sum_{i=1}^{S} p_i p_i^{q-1}}}$$

The denominator equals mean proportional species abundance in the dataset as calculated with the weighted generalized mean with exponent q - 1. In the equation, S is the total number of species (species richness) in the dataset, and the proportional abundance of the ith species is p_i The proportional abundances themselves are used as weights. The equation is often written in the equivalent form:

$$^{q}D = \left(\sum_{i=1}^{S} p_i^q \right)^{1/(1-q)}$$

The value of q defines which kind of mean is used. $q = 0$ corresponds to the weighted harmonic mean, which is 1/S because the p_i values cancel out. $q = 1$ is undefined, except that the limit as q approaches 1 is well defined:

$$\lim_{q \to 1} {}^{q}D = \exp\left(-\sum_{i=1}^{S} p_i \ln p_i \right)$$

$q = 2$ corresponds to the arithmetic mean. As q approaches infinity, the generalized mean approaches the maximum p_i value. In practice, q modifies species weighting, such that increasing q increases the weight given to the most abundant species, and fewer equally abundant species are hence needed to reach mean proportional abundance. Consequently, large values of q lead to smaller species diversity than small values of q for the same dataset. If all species are equally abundant in the dataset, changing the value of q has no effect, but species diversity at any value of q equals species richness.

Negative values of q are not used, because then the effective number of species (diversity) would exceed the actual number of species (richness). As q approaches negative infinity, the generalized mean approaches the minimum p_i value. In many real datasets, the least abundant species is represented by a single individual, and then the effective number of species would equal the number of individuals in the dataset.

The same equation can be used to calculate the diversity in relation to any classification, not only species. If the individuals are classified into genera or functional types, p_i represents the proportional abundance of the ith genus or functional type, and ^{q}D equals genus diversity or functional type diversity, respectively.

Diversity Indices

Often researchers have used the values given by one or more diversity indices to quantify species diversity. Such indices include species richness, the Shannon index, the Simpson index, and the complement of the Simpson index (also known as the Gini-Simpson index).

When interpreted in ecological terms, each one of these indices corresponds to a different thing, and their values are therefore not directly comparable. Species richness quantifies the actual rather than effective number of species. The Shannon index equals $\log({}^qD)$, and in practice quantifies the uncertainty in the species identity of an individual that is taken at random from the dataset. The Simpson index equals $1/{}^qD$ and quantifies the probability that two individuals taken at random from the dataset (with replacement of the first individual before taking the second) represent the same species. The Gini-Simpson index equals $1 - 1/{}^qD$ and quantifies the probability that the two randomly taken individuals represent different species.

Sampling Considerations

Depending on the purposes of quantifying species diversity, the dataset used for the calculations can be obtained in different ways. Although species diversity can be calculated for any dataset where individuals have been identified to species, meaningful ecological interpretations require that the dataset is appropriate for the questions at hand. In practice, the interest is usually in the species diversity of areas so large that not all individuals in them can be observed and identified to species, but a sample of the relevant individuals has to be obtained. Extrapolation from the sample to the underlying population of interest is not straightforward, because the species diversity of the available sample generally gives an underestimation of the species diversity in the entire population. Applying different sampling methods will lead to different sets of individuals being observed for the same area of interest, and the species diversity of each set may be different. When a new individual is added to a dataset, it may introduce a species that was not yet represented. How much this increases species diversity depends on the value of q: when $q = 0$, each new actual species causes species diversity to increase by one effective species, but when q is large, adding a rare species to a dataset has little effect on its species diversity.

In general, sets with many individuals can be expected to have higher species diversity than sets with fewer individuals. When species diversity values are compared among sets, sampling efforts need to be standardised in an appropriate way for the comparisons to yield ecologically meaningful results. Resampling methods can be used to bring samples of different sizes to a common footing. Species accumulation curves and the number of species only represented by one or a few individuals can be used to help in estimating how representative the available sample is of the population from which it was drawn.

Trends in Species Diversity

The observed species diversity is affected not only by the number of individuals but also by the heterogeneity of the sample. If individuals are drawn from different environmental conditions (or different habitats), the species diversity of the resulting set can be expected to be higher than if all individuals are drawn from a similar environment. Increasing the area sampled increases observed species diversity both because more individuals get included in the sample and because large areas are environmentally more heterogeneous than small areas.

Productivity and Stability as Indicators of Ecosystem Health

The importance of stability in community ecology is clear. An unstable ecosystem will be more likely to lose species. Thus, if there is indeed a link between diversity and stability, it is likely that losses of diversity could feedback on themselves, causing even more losses of species. Productivity, on the other hand, has a less clear importance in community ecology. In managed areas like cropland, and in areas where animals are grown or caught, increasing productivity increases the economic success of the area and implies that the area has become more efficient, leading to possible long term resource sustainability. It is more difficult to find the importance of productivity in natural ecosystems.

Beyond the value biodiversity has in regulating and stabilizing ecosystem processes, there are direct economic consequences of losing diversity in certain ecosystems and in the world as a whole. Losing species means losing potential food, medicines, industrial products, and tourism, all of which have a direct economic effect on people's lives.

Effects on Community Productivity

- **Complementarity:** Plant species coexistence is thought to be the result of niche partitioning, or differences in resource requirements among species. By complementarity, a more diverse plant community should be able to use resources more completely, and thus be more productive. Also called niche differentiation, this mechanism is a central principle in the functional group approach, which breaks species diversity down into functional components.
- **Facilitation:** Facilitation is a mechanism whereby certain species help or allow other species to grow by modifying the environment in a way that is favorable to a co-occurring species. Plants can interact through an intermediary like nitrogen, water, temperature, space, or interactions with weeds or herbivores among others. Some examples of facilitation include large desert perennials acting as nurse plants, aiding the establishment of young neighbors of other species by alleviating water and temperature stress, and nutrient enrichment by nitrogen-fixers such as legumes.
- **The Sampling Effect:** The sampling effect of diversity can be thought of as having a greater chance of including a species of greatest inherent productivity in a plot that is more diverse. This provides for a composition effect on productivity, rather than diversity being a direct cause. However, the sampling effect may in fact be a compilation of different effects. The sampling effect can be separated into the greater likelihood of selecting a species that is, 1) adapted well to particular site conditions, or 2) of a greater inherent productivity. Additionally,

one can add to the sampling effect a greater likelihood of including, 3) a pair of species that highly complement each other, or 4) a certain species with a large facilitative effect on other members of the community.

Review of Data

Field experiments to test the degree to which diversity affects community productivity have had variable results, but many long term studies in grassland ecosystems have found that diversity does indeed enhance the productivity of ecosystems. Additionally, evidence of this relationship has also been found in grassland microcosms. The differing results between studies may partially be attributable to their reliance on samples with equal species diversities rather than species diversities that mirror those observed in the environment. A 2006 experiment utilizing a realistic variation in species composition for its grassland samples found a positive correlation between increased diversity and increased production.

However, these studies have come to different conclusions as to whether the cause was due more to diversity or to species composition. Specifically, a diversity in the functional roles of the species may be a more important quality for predicting productivity than the diversity in species number. Recent mathematical models have highlighted the importance of ecological context in unraveling this problem. Some models have indicated the importance of disturbance rates and spatial heterogeneity of the environment, others have indicated that the time since disturbance and the habitat's carrying capacity can cause differing relationships. Each ecological context should yield not only a different relationship, but a different contribution to the relationship due to diversity and to composition. The current consensus holds at least that certain combinations of species provide increased community productivity.

Future Research

In order to correctly identify the consequences of diversity on productivity and other ecosystem processes, many things must happen. First, it is imperative that scientists stop looking for a single relationship. It is obvious now from the models, the data, and the theory that there is no one overarching effect of diversity on productivity. Scientists must try to quantify the differences between composition effect and diversity effects, as many experiments never quantify the final realized species diversity (instead only counting numbers of species of seeds planted) and confound a sampling effect for facilitators (a compositional factor) with diversity effects.

Relative amounts of overyielding (or how much more a species grows when grown with other species than it does in monoculture) should be used rather than absolute amounts as relative overyielding can give clues as to the mechanism by which diversity is influencing productivity. However if experimental protocols are incomplete, one may be able to indicate the existence of a complementary or facilitative effect in the experiment, but not be able to recognize its cause. Experimenters should know what the goal of their experiment is, that is, whether it is meant to inform natural or managed ecosystems, as the sampling effect may only be a real effect of diversity in natural ecosystems (managed ecosystems are composed to maximize complementarity and facilitation regardless of species number). By knowing this, they should be able to choose spatial and temporal scales that are appropriate for their experiment. Lastly, to resolve the diversity-function debate, it is advisable that experiments be done with large amounts of spatial and resource heterogeneity and environmental fluctuation over time, as these types of experiments should be able to demonstrate the diversity-function relationship more easily.

Effects on Community Stability

- **Averaging Effect:** If all species have differential responses to changes in the ecosystem over time, then the averaging of these responses will cause a more temporally stable ecosystem if more species are in the ecosystem. This effect is a statistical effect due to summing random variables.
- **Negative Covariance Effect:** If some species do better when other species are not doing well, then when there are more species in the ecosystem, their overall variance will be lower than if there were fewer species in the system. This lower variance indicates higher stability. This effect is a consequence of competition as highly competitive species will negatively covary.
- **Insurance Effect:** If an ecosystem contains more species then it will have a greater likelihood of having redundant stabilizing species, and it will have a greater number of species that respond to perturbations in different ways. This will enhance an ecosystem's ability to buffer perturbations.
- **Resistance to Invasion:** Diverse communities may use resources more completely than simple communities because of a diversity effect for complementarity. Thus invaders may have reduced success in diverse ecosystems, or there may be a reduced likelihood that an invading species will introduce a new property or process to a diverse ecosystem.
- **Resistance to Disease:** A decreased number of competing plant species may allow the abundances of other species to increase, facilitating the spread of diseases of those species.

Review of Temporal Stability Data

Models have predicted that empirical relationships between temporal variation of community productivity and species diversity are indeed real, and that they almost have to be. Some temporal stability data can be almost completely explained by the averaging effect by constructing null models to test the data against. Competition, which causes negative covariances, only serves to strengthen these relationships.

Review of Resistance and Resilience Stability Data

This area is more contentious than the area of temporal stability, mostly because some have tried generalizing the findings of the temporal stability models and theory to stability in general. While the relationship between temporal variations in productivity and diversity has a mathematical cause, which will allow the relationship to be seen much more often than not, it is not the case with resistance/resilience stability. Some experimenters have seen a correlation between diversity and reduced invasibility, though many have also seen the opposite. The correlation between diversity and disease is also tenuous, though theory and data do seem to support it.

Future Research

In order to more fully understand the effects of diversity on the temporal stability of ecosystems it is necessary to recognize that they are bound to occur. By constructing null models to test the data against (as in Doak et al. 1998) it becomes possible to find situations and ecological contexts where ecosystems become more or less stable than they should be. Finding these contexts would allow for mechanistic studies into why these ecosystems are more stable, which may allow for applications in conservation management.

More importantly, more complete experiments into whether diverse ecosystems actually resist invasion and disease better than their less diverse equivalents as invasion and disease are two important factors that lead to species extinctions in the present day.

Theory and Preliminary Effects from Examining Food Webs

One major problem with both the diversity-productivity and diversity-stability debates discussed up to this point is that both focus on interactions at just a single trophic level. That is, they are concerned with only one level of the food web, namely plants. Other research, unconcerned with the effects of diversity, has demonstrated strong top-down forcing of ecosystems (see keystone species). There is little actual data available regarding the effects of different food webs, but theory helps us in this area. First, if a food web in an ecosystem has a lot of weak interactions between different species, then it should have more stable populations and the community as a whole should be more stable. If upper levels of the web are more diverse, then there will be less biomass in the lower levels and if lower levels are more diverse they will better be able to resist consumption and be more stable in the face of consumption. Also, top-down forcing should be reduced in less diverse ecosystems because of the bias for species in higher trophic levels to go extinct first. Lastly, it has recently been shown that consumers can dramatically change the biodiversity-productivity-stability relationships that are implied by plants alone. Thus, it will be important in the future to incorporate food web theory into the future study of the effects of biodiversity. In addition this complexity will need to be addressed when designing biodiversity management plans.

REFERENCES

1. "National Biological Information Infrastructure". Introduction to Genetic Diversity. U.S. Geological Survey. Archived from the original on February 25, 2011. Retrieved March 1, 2011.

2. Study: Loss Of Genetic Diversity Threatens Species Diversity
3. Nevo, Eviatar (May 2001). "Evolution of Genome-Phenome Diversity under Environmental Stress". Proceedings of the National Academy of Sciences of the United States of America. 98 (11): 6233–6240. doi:10.1073/pnas.101109298. JSTOR 3055788. PMC 33451 . PMID 11371642.
4. Groom, M.J., Meffe, G.K. and Carroll, C.R. (2006) Principles of Conservation Biology (3rd ed.). Sunderland, MA: Sinauer Associates. Website with additional information: http://www.sinauer.com/groom/
5. Tisdell, C. (2003). "Socioeconomic causes of loss of animal genetic diversity: analysis and assessment". Ecological Economics. 45 (3): 365–376. doi:10.1016/S0921-8009(03)00091-0.
6. Frankham, Richard (November 2005). "Genetics and Extinction". Biological Conservation. 126 (2): 131–140. doi:10.1016/j.biocon.2005.05.002. Retrieved 2012-10-24. The rate of evolutionary change (R) is determined primarily by the quantitative genetic variation
7. Pullin, Andrew S. (2002). Conservation biology (1st publ. ed.). Cambridge: Cambridge University Press. ISBN 9780521644822.
8. " Genetic Diversity." National Biological Information Infrastructure. NBII. 16 Mar. 2008 www.nbii.gov
9. Paetkau, David; Waits, Lisette P.; Clarkson, Peter L.; Craighead, Lance; Vyse, Ernie; Ward, Ryk; Strobeck, Curtis (2008). "Variation in Genetic Diversity across the Range of North American Brown Bears". Conservation Biology. 12 (2): 418. doi:10.1111/j.1523-1739.1998.96457.x.
10. "Introduction to Genetic Diversity." Cheetah Conservation Fund. 2002. 19 Mar. 2008 www.cheetah.org
11. "Monoculture and the Irish Potato Famine: cases of missing genetic variation". University of California Museum of Paleontology, Berkeley, California. Retrieved 2014-03-28.
12. FAO. 2015. The Second Report on the State of the World's Animal Genetic Resources for Food and Agriculture. Rome.
13. FAO. (2015). The Second Report on the State of the World's Animal Genetic Resources for Food and Agriculture. Rome.
14. "Scientists Discover Interplay Between Genes and Viruses in Tiny Ocean Plankton". National Science Foundation. March 23, 2006. Retrieved December 12, 2008.
15. Stephens, Tim. "Currents." University of California, Santa Cruz. 10 Aug. 1998. University of California. 19 Mar. 2008 www.ucsc.edu
16. Fildes, Jonathan (May 29, 2007). "Cheating cheetahs caught by DNA". BBC News. Retrieved December 12, 2008.
17. Kawabe, K.; Worawut, R.; Taura, S.; Shimogiri, T.; Nishida, T.; Okamoto, S. (2014-01-01). "Genetic Diversity of mtDNA D-loop Polymorphisms in Laotian Native Fowl Populations". Asian-Australasian Journal of Animal Sciences. 27

(1): 19–23. doi:10.5713/ajas.2013.13443. ISSN 1011-2367. PMC 4093284 . PMID 25049921.

18. Sahney, S.; Benton, M.J. & Ferry, P.A. (2010). "Links between global taxonomic diversity, ecological diversity and the expansion of vertebrates on land". Biology Letters. 6 (4): 544–547. doi:10.1098/rsbl.2009.1024. PMC 2936204 . PMID 20106856.
19. Vitousek, P.M., Mooney, H. A., and Lubchenco, J. et al. 1997.Human domination of Earth's ecosystems. Science 277: 494-499.
20. Hines, J., W. H. van der Putten, G. B. De Deyn, C. Wagg, W. Voigt, C. Mulder, W. Weisser, J. Engel, C. Melian, S. Scheu, K. Birkhofer, A. Ebeling, C. Scherber, N. Eisenhauer. 2015. Towards an integration of biodiversity-ecosystem functioning and food-web theory to evaluate connections between multiple ecosystem services. In Guy Woodward, David A. Bohan, editors: Ecosystem Services: From Biodiversity to Society, Part 1, Vol 53, Advances in Ecological Research, UK: Academic Press, pp. 161-199
21. Doak, D.F., Bigger, D., Harding, E.K., et al. 1998. The statistical inevitability of stability-diversity relationships in community ecology. Am. Nat. 151: 264-276.
22. McCann, K.S. 2000. The diversity-stability debating. Nature 405: 228-233.
23. Fridley, J.D. 2001. The influence of species diversity on ecosystem productivity: how, where, why? Oikos 93: 514-526.
24. Wilson,E.O. 1992. The Diversity of Life. Harvard Univ. Press, Cambridge, Mass., U.S.A.
25. Tilman, D., Knops, J., Wedin, D. et al. 1997a. The influence of functional diversity and composition on ecosystem processes. Science 277: 1300-1302.
26. Tilman, D., Lehman, C.L. and Thomson, K.T. 1997b. Plant diversity and ecosystem productivity: theoretical considerations. Proc. Natl. Acad. Sci. USA 94: 1857-1861.
27. Tilman, D. 1999. The ecological consequences of changes in biodiversity: a search for general principles. Ecology 80: 1455-1474.
28. Vandermeer, J. H. 1989. The ecology of intercropping. Cambridge Univ. Press., Cambridge, England.
29. Turner, R.M., Alcorn, S.M., Olin, G. and Booth, J.A. 1966. The influence of shade, soil, and water on saguaro seedling establishment.Bot. Gaz. 127: 95-102.
30. Tilman, D., Wedin, D, and Knops, J. 1996. Productivity and sustainability influenced by biodiversity in grassland ecosystems. Nature 379: 718-720.
31. Naeem, S., Thompson, L.J., Lawler, S.P, et al. 1994. Declining biodiversity can alter the performance of ecosystems. Nature 368: 734-737.
32. Hooper, D. and Vitousek, P. 1997. The effect of plant composition and diversity on ecosystem processes. Science 277: 1302-1305.
33. Zavaleta, E. S.; Hulvey, K. B. (2006). "Realistic variation in species composition affects grassland production, resource use and invasion resistance" (PDF). Plant Ecology. 188: 39. doi:10.1007/s11258-006-9146-z. Retrieved 18 January 2014.

34. Cardinale B.J., Nelson K., Palmer M.A. 2000. Linking species diversity to the functioning of ecosystems: on the importance of environmental context. Oikos 91: 175-183.

35. Aarssen, L.W., Laird, R.A., and Pither J. 2003. Is the productivity of vegetation plots higher or lower when there are more species? Variable predictions from interaction of the "sampling effect" and "competitive dominance effect" on the habitat templet. Oikos 102: 427-432.

36. Hooper, D. U.; Chapin, F. S.; Ewel, J. J.; Hector, A.; Inchausti, P.; Lavorel, S.; Lawton, J. H.; Lodge, D. M.; Loreau, M.; Naeem, S.; Schmid, B.; Setälä, H.; Symstad, A. J.; Vandermeer, J.; Wardle, D. A. (2005). "EFFECTS OF BIODIVERSITY ON ECOSYSTEM FUNCTIONING: A CONSENSUS OF CURRENT KNOWLEDGE". Ecological Monographs. 75: 3. doi:10.1890/04-0922. Retrieved 18 January 2014.

37. Tilman, D., Lehman, C.L. and Bristow, C.E. 1998. Diversity-stability relationships: statistical inevitability or ecological consequence. Am. Nat. 151: 264-276.

38. Naeem, S. and Li, S. 1997. Biodiversity enhances ecosystem reliability. Nature 390: 507-509.

39. Elton, C.S. 1958. The ecology of invasions by animals and plants. John Wiley, New York, New York, USA.

40. Chapin, F.S., III, Walker, B.H., Hobbs, R.J. et al. 1997. Biotic control over the functioning of ecosystems. Science 277: 500-504.

41. Mitchell, C.E., Tilman, D. and Groth, J.V., Effects of grassland plant species diversity, abundance, and composition on foliar fungal disease. Ecology 83: 1713-1726.

42. Dukes, J.S. 2001. Biodiversity and invisibility in grassland microcosms. Oecologia 126: 563-568.

43. Duffy, J.E. 2002. Biodiversity and ecosystem function: the consumer connection. Oikos99: 201-219

44. Worm, B. and Duffy, J.E. 2003.Biodiversity, productivity and stability in real food webs. Trends in Ecology and Evolution 18: 628-632.

45. Hill, M. O. (1973) Diversity and evenness: a unifying notation and its consequences. Ecology, 54, 427–432

46. Tuomisto, H. (2010) A diversity of beta diversities: straightening up a concept gone awry. Part 1. Defining beta diversity as a function of alpha and gamma diversity. Ecography, 33, 2-22. doi:10.1111/j.1600-0587.2009.05880.x

47. Tuomisto, H. 2010. A consistent terminology for quantifying species diversity? Yes, it does exist. Oecologia 4: 853–860. doi:10.1007/s00442-010-1812-0

48. Krebs, C. J. (1999) Ecological Methodology. Second edition. Addison-Wesley, California.

49. Magurran, A. E. (2004) Measuring biological diversity. Blackwell Publishing, Oxford.

50. Jost, L. (2006) Entropy and diversity. Oikos, 113, 363–375

51. Jost, L. (2007) Partitioning diversity into independent alpha and beta components. Ecology, 88, 2427–2439.

52. Tuomisto, H. (2010) A diversity of beta diversities: straightening up a concept gone awry. Part 2. Quantifying beta diversity and related phenomena. Ecography, 33, 23-45. doi:10.1111/j.1600-0587.2009.06148.x

53. Colwell, R. K. and Coddington, J. A. (1994) Estimating terrestrial biodiversity through extrapolation. Philosophical Transactions: Biological Sciences, 345, 101-118.

54. Good, I. J. and Toulmin, G. H. (1956) The number of new species, and the increase in population coverage, when a sample is increased. Biometrika, 43, 45-63.

55. Chao, A. (2005) Species richness estimation. Pages 7909-7916 in N. Balakrishnan, C. B. Read, and B. Vidakovic, eds. Encyclopedia of Statistical Sciences. New York, Wiley.

CHAPTER 5

Habitat Destruction

Chapter Contents

A habitat is the natural place where plants, animals, or other organisms live; it is what they call home. This is the living area necessary for an ecosystem to remain healthy. Destruction means to change something so much that it can no longer exist as it once was. So, when we put the two definitions together we get: habitat + destruction = a home to species that has been changed to the point it no longer exists.

INTRODUCTION

Habitat destruction is the process in which natural habitat is rendered unable to support the species present. In this process, the organisms that previously used the site are displaced or destroyed, reducing biodiversity. Habitat destruction by human activity is mainly for the purpose of harvesting natural resources for industry production and urbanization. Clearing habitats for agriculture is the principal cause of habitat destruction. Other important causes of habitat destruction include mining, logging, trawling, and urban sprawl. Habitat destruction is currently ranked as the primary cause of species extinction worldwide. It is a process of natural environmental change that may be caused by habitat fragmentation, geological processes, climate change, or by human activities such as the introduction of invasive species, ecosystem nutrient depletion, and other human activities

The terms habitat loss and habitat reduction are also used in a wider sense, including loss of habitat from other factors, such as water and noise pollution.

HABITAT

A habitat is an ecological or environmental area that is inhabited by a particular species of animal, plant, or other type of organism. The term typically refers to the zone in which the organism lives and where it can find food, shelter, protection, and mates for reproduction. It is the natural environment in which an organism lives, or the physical environment that surrounds a species population.

A habitat is made up of physical factors such as soil, moisture, range of temperature, and light intensity as well as biotic factors such as the availability of food and the presence or absence of predators. Every organism has certain habitat needs for the conditions in which it will thrive, but some are tolerant of wide variations while others are very specific in their requirements. A habitat is not necessarily a geographical area, it can be the interior of a stem, a rotten log, a rock or a clump of moss, and for a parasitic organism it is the body of its host, part of the host's body such as the digestive tract, or a single cell within the host's body.

Habitat types include polar, temperate, subtropical and tropical. The terrestrial vegetation type may be forest, steppe, grassland, semi-arid, or desert. Fresh water habitats include marshes, streams, rivers, lakes, ponds and estuaries, and marine habitats include salt marshes, the coast, the intertidal zone, reefs, bays, the open sea, the sea bed, deep water and submarine vents.

Habitats change over time. This may be due to a violent event such as the eruption of a volcano, an earthquake, a tsunami, a wildfire, or a change in oceanic currents; or the change may be more gradual over millennia with alterations in the climate, as ice sheets and glaciers advance and retreat, and as different weather patterns bring changes of precipitation and solar radiation. Other changes come as a direct result of human activities; deforestation, the ploughing of ancient grasslands, the diversion and damming of rivers, the draining of marshland, and the dredging of the seabed. The introduction of alien species can have a devastating effect on native wildlife, through increased predation, competition for resources or the introduction of pests and diseases to which the native species have no immunity.

This coral reef in the Phoenix Islands Protected Area is a rich habitat for sea life.

Few creatures make the ice shelves of Antarctica their habitat.

Ibex in alpine habitat.

Definition and Etymology

The word "habitat" has been in use since about 1755 and derives from the Latin third-person singular present indicative of *habitāre*, to inhabit, from *habēre*, to have or to hold. Habitat can be defined as the natural environment of an organism, the place in which it is natural for it to live and grow. It is similar in meaning to a biotope, an area of uniform environmental conditions associated with a particular community of plants and animals.

Environmental Factors

The chief environmental factors affecting the distribution of living organisms are temperature, humidity, climate, soil type and light intensity, and the presence or absence of all the requirements that the organism needs to sustain it. Generally speaking, animal communities are reliant on specific types of plant communities.

Some plants and animals are generalists, and their habitat requirements are met in a wide range of locations. The small white butterfly (*Pieris rapae*) for example is found on all the continents of the world apart from Antarctica. Its larvae feed on a wide range of *Brassicas* and various other plant species, and it thrives in any open location with diverse plant associations. The large blue butterfly is much more specific in its requirements; it is found only in chalk grassland areas, its larvae feed on *Thymus* species and because of complex lifecycle requirements it inhabits only areas in which *Myrmica* ants live.

Disturbance is important in the creation of biodiverse habitats. In the absence of disturbance, a climax vegetation cover develops that prevents the establishment of other species. Wildflower meadows are sometimes created by conservationists but most of the flowering plants used are either annuals or biennials and disappear after a few years in the absence of patches of bare ground on which their seedlings can grow. Lightning strikes and toppled trees in tropical forests allow species richness to be maintained as pioneering species move in to fill the gaps created. Similarly coastal habitats can become dominated by kelp until the seabed is disturbed by a storm and the algae swept away, or shifting sediment exposes new areas for colonisation. Another cause of disturbance is when an area may be overwhelmed by an invasive introduced species which is not kept under control by natural enemies in its new habitat.

Types of Habitat

Terrestrial habitat types include forests, grasslands, wetlands, and deserts. Within these broad biomes are more specific habitats with varying climate types, temperature regimes, soils, altitudes and vegetation types. Many of these habitats grade into each other and each one has its own typical communities of plants and animals. A habitat may suit a particular species well, but its presence or absence at any particular location depends to some extent on chance, on its dispersal abilities and its efficiency as a coloniser.

Rich rainforest habitat in Dominica.

Wetland habitats in Borneo.

Freshwater habitats include rivers, streams, lakes, ponds, marshes, and bogs. Although some organisms are found across most of these habitats, the majority have more specific requirements. The water velocity, its temperature and oxygen saturation are important factors, but in river systems, there are fast and slow sections, pools, bayous and backwaters which provide a range of habitats. Similarly, aquatic plants can be floating, semi-submerged, submerged or grow in permanently or temporarily saturated soils besides bodies of water. Marginal plants provide important habitat for both invertebrates and vertebrates, and submerged plants provide oxygenation of the water, absorb nutrients and play a part in the reduction of pollution.

Marine habitats include brackish water, estuaries, bays, the open sea, the intertidal zone, the sea bed, reefs and deep water zones. Further variations include rock pools, sand banks, mudflats, brackish lagoons, sandy and pebbly beaches, and seagrass beds, all supporting their own flora and fauna. The benthic zone or seabed provides a home for both static organisms, anchored to the substrate, and for a large range of organisms crawling on or burrowing into the surface. Some creatures float among the waves on the surface of the water, or raft on floating debris, others swim at a range of depths, including organisms in the demersal zone close to the seabed, and myriads of organisms drift with the currents and form the plankton.

Desert scene in Egypt

A desert is not the kind of habitat that favours the presence of amphibians, with their requirement for water to keep their skins moist and for the development of their young. Nevertheless, some frogs live in deserts, creating moist habitats underground and hibernating while conditions are adverse. Couch's spadefoot toad (*Scaphiopus couchii*) emerges from its burrow when a downpour occurs and lays its eggs in the transient pools that form; the tadpoles develop with great rapidity, sometimes in as little as nine days, undergo metamorphosis, and feed voraciously before digging a burrow of their own.

Other organisms cope with the drying up of their aqueous habitat in other ways. Vernal pools are ephemeral ponds that form in the rainy season and dry up afterwards.

They have their specially-adapted characteristic flora, mainly consisting of annuals, the seeds of which survive the drought, but also some uniquely adapted perennials. Animals adapted to these extreme habitats also exist; fairy shrimps can lay "winter eggs" which are resistant to desiccation, sometimes being blown about with the dust, ending up in new depressions in the ground. These can survive in a dormant state for as long as fifteen years. Some killifish behave in a similar way; their eggs hatch and the juvenile fish grow with great rapidity when the conditions are right, but the whole population of fish may end up as eggs in diapause in the dried up mud that was once a pond.

Many animals and plants have taken up residence in urban environments. They tend to be adaptable generalists and use the town's features to make their homes. Rats and mice have followed man around the globe, pigeons, peregrines, sparrows, swallows and house martins use the buildings for nesting, bats use roof space for roosting, foxes visit the garbage bins, and squirrels, coyotes, raccoons and skunks roam the streets. About 2,000 coyotes are thought to live in and around Chicago. A survey of dwelling houses in northern European cities in the twentieth century found about 175 species of invertebrate inside them, including 53 species of beetle, 21 flies, 13 butterflies and moths, 13 mites, 9 lice, 7 bees, 5 wasps, 5 cockroaches, 5 spiders, 4 ants, and a number of other groups. In warmer climates, termites are serious pests in the urban habitat; 183 species are known to affect buildings and 83 species cause serious structural damage.

Microhabitats

A microhabitat is the small-scale physical requirements of a particular organism or population. Every habitat includes large numbers of microhabitats with subtly different exposure to light, humidity, temperature, air movement, and other factors. The lichens that grow on the north face of a boulder are different to those that grow on the south face, from those on the level top and those that grow on the ground nearby; the lichens growing in the grooves and on the raised surfaces are different from those growing on the veins of quartz. Lurking among these miniature "forests" are the microfauna, each species of invertebrate with its own specific habitat requirements.

There are numerous different microhabitats in a wood; coniferous forest, broad-leafed forest, open woodland, scattered trees, woodland verges, clearings and glades; tree trunk, branch, twig, bud, leaf, flower and fruit; rough bark, smooth bark, damaged bark, rotten wood, hollow, groove and hole; canopy, shrub layer, plant layer, leaf litter and soil; buttress root, stump, fallen log, stem base, grass tussock, fungus, fern and moss. The greater the structural diversity in the wood, the greater the number of microhabitats that will be present. A range of tree species with individual specimens of varying sizes and ages, and a range of features such as streams, level areas, slopes, tracks, clearings and felled areas will provide suitable conditions for an enormous number of biodiverse plants and animals. For example, in Britain it has been estimated that various types of rotting wood are home to over 1,700 species of invertebrate.

For a parasitic organism, its habitat is the particular part of the outside or inside of its host on or in which it is adapted to live. The life cycle of some parasites involves several different host species, as well as free-living life stages, sometimes providing vastly different microhabitats. One such organism is the trematode (flatworm) *Microphallus turgidus*, present in brackish water marshes in the southeastern United States. Its first intermediate host is a snail and the second, a glass shrimp. The final host is the waterfowl or mammal that consumes the shrimp.

Extreme Habitats

An Antarctic rock split apart to show an endolithic lifeform showing as a green layer a few millimetres thick.

Although the vast majority of life on Earth lives in mesophyllic (moderate) environments, a few organisms, most of them microbes, have managed to colonise extreme environments that are unsuitable for most higher life forms. There are bacteria, for example, living in Lake Whillans, half a mile below the ice of Antarctica; in the absence of sunlight, they must rely on organic material from elsewhere, perhaps decaying matter from glacier melt water or minerals from the underlying rock. Other bacteria can be found in abundance in the Mariana Trench, the deepest place in the ocean and on Earth; marine snow drifts down from the surface layers of the sea and accumulates in this undersea valley, providing nourishment for an extensive community of bacteria.

Other microbes live in habitats lacking in oxygen, and are dependent on chemical reactions other than photosynthesis. Boreholes drilled 300 m (1,000 ft) into the rocky seabed have found microbial communities apparently based on the products of reactions between water and the constituents of rocks. These communities have been little studied, but may be an important part of the global carbon cycle. Rock in mines two miles deep also harbour microbes; these live on minute traces of hydrogen produced in slow oxidizing reactions inside the rock. These metabolic reactions allow life to exist in places with no oxygen or light, an environment that had previously been thought to be devoid of life.

The intertidal zone and the photic zone in the oceans are relatively familiar habitats. However the vast bulk of the ocean is unhospitable to air-breathing humans, with scuba divers limited to the upper 50 m (160 ft) or so. The lower limit for photosynthesis is 100 to 200 m (330 to 660 ft) and below that depth the prevailing conditions include total darkness, high pressure, little oxygen (in some places), scarce food resources and extreme cold. This habitat is very challenging to research, and as well as being little studied, it is vast, with 79% of the Earth's biosphere being at depths greater than 1,000 m (3,300 ft). With no plant life, the animals in this zone are either detritivores, reliant on food drifting down from surface layers, or they are predators, feeding on each other. Some organisms are pelagic, swimming or drifting in mid-ocean, while others are benthic, living on or near the seabed. Their growth rates and metabolisms tend to be slow, their eyes may be very large to detect what little illumination there is, or they may be blind and rely on other sensory inputs. A number of deep sea creatures are bioluminescent; this serves a variety of functions including predation, protection and social recognition. In general, the bodies of animals living at great depths are adapted to high pressure environments by having pressure-resistant biomolecules and small organic molecules present in their cells know as piezolytes, which give the proteins the flexibility they need. There are also unsaturated fats in their membranes which prevent them from solidifying at low temperatures.

Dense mass of white crabs at a hydrothermal vent, with stalked barnacles on right.

Hydrothermal vents were first discovered in the ocean depths in 1977. They resulted from seawater becoming heated after seeping through cracks to places where hot magma is close to the seabed. The under water hot springs may gush forth at temperatures of over 340 °C (640 °F) and support unique communities of organisms in their immediate vicinity. The basis for this teeming life is chemosynthesis, a process by which microbes convert such substances as hydrogen sulfide or ammonia into organic molecules. These bacteria and Archaea are the primary producers in these ecosystems and support a diverse array of life. About 350 species of organism, dominated by molluscs, polychaete worms and crustaceans, had been discovered around hydrothermal vents by the end of the twentieth century, most of them being new to science and endemic to these habitats.

Besides providing locomotion opportunities for winged animals and a conduit for the dispersal of pollen grains, spores and seeds, the atmosphere can be considered to be a habitat in its own right. There are metabolically active microbes present that actively reproduce and spend their whole existence airborne, with hundreds of thousands of individual organisms estimated to be present in a cubic metre of air. The airborne microbial community may be as diverse as that found in soil or other terrestrial environments, however these organisms are not evenly distributed, their densities varying spatially with altitude and environmental conditions. Aerobiology has been little studied, but there is evidence of nitrogen fixation in clouds, and less clear evidence of carbon cycling, both facilitated by microbial activity.

There are other examples of extreme habitats where specially adapted lifeforms exist; tar pits teeming with microbial life; naturally occurring crude oil pools inhabited by the larvae of the petroleum fly; hot springs where the temperature may be as high as 71 °C (160 °F) and cyanobacteria create microbial mats; cold seeps where the methane and hydrogen sulfide issue from the ocean floor and support microbes and higher animals such as mussels which form symbiotic associations with these anaerobic organisms; salt pans harbour salt-tolerant microorganisms and also *Wallemia ichthyophaga*, a basidomycotous fungus; ice sheets in Antarctica which support fungi *Thelebolus* spp., and snowfields on which algae grow.

Habitat Change

Twenty five years after the devastating eruption at Mount St. Helens, United States, pioneer species have moved in.

Whether from natural processes or the activities of man, landscapes and their associated habitats change over time. There are the slow geomorphological changes associated with the geologic processes that cause tectonic uplift and subsidence, and the more rapid changes associated with earthquakes, landslides, storms, flooding, wildfires, coastal erosion, deforestation, and changes in land use. Then there are the changes in habitats brought on by alterations in farming practices, tourism, pollution, fragmentation, and climate change.

Loss of habitat is the single greatest threat to any species. If an island on which an endemic organism lives becomes uninhabitable for some reason, the species will become extinct. Any type of habitat surrounded by a different habitat is in a similar situation to an island. If a forest is divided into parts by logging, with strips of cleared land separating woodland blocks, and the distances between the remaining fragments exceeds the distance an individual animal is able to travel, that species becomes especially vulnerable. Small populations generally lack genetic diversity and may be threatened by increased predation, increased competition, disease and unexpected catastrophe. At the edge of each forest fragment, increased light encourages secondary growth of fast-growing species and old growth trees are more vulnerable to logging as access is improved. The birds that nest in their crevices, the epiphytes that hang from their branches and the invertebrates in the leaf litter are all adversely affected and biodiversity is reduced. Habitat fragmentation can be ameliorated to some extent by the provision of wildlife corridors connecting the fragments. These can be a river, ditch, strip of trees, hedgerow or even an underpass to a highway. Without the corridors, seeds cannot disperse and animals, especially small ones, cannot travel through the hostile territory, putting populations at greater risk of local extinction.

Habitat disturbance can have long-lasting effects on the environment. *Bromus tectorum* is a vigorous grass from Europe which has been introduced to the United States where it has become invasive. It is highly adapted to fire, producing large amounts of flammable detritus and increasing the frequency and intensity of wildfires. In areas where it has become established, it has altered the local fire regimen to such an extant that native plants cannot survive the frequent fires, allowing it to become even more dominant. A marine example is when sea urchin populations "explode" in coastal waters and destroy all the macroalgae present. What was previously a kelp forest becomes an urchin barren that may last for years and this can have a profound effect on the food chain. Removal of the sea urchins, by disease for example, can result in the seaweed returning, with an over-abundance of fast-growing kelp.

Habitat Protection

The protection of habitats is a necessary step in the maintenance of biodiversity because if habitat destruction occurs, the animals and plants reliant on that habitat will suffer. Many countries have enacted legislation to protect their wildlife. This may take the form of the setting up of national parks, forest reserves and wildlife reserves, or it may restrict the activities of humans with the objective of benefiting wildlife. The laws may be designed to protect a particular species or group of species, or the legislation may prohibit such activities as the collecting of bird eggs, the hunting of animals, or the removal of plants. A general law on the protection of habitats may be more difficult to implement than a site specific requirement. A concept introduced in the Unites States in 1973 involves protecting the critical habitat of endangered species, and a similar concept has been incorporated into some Australian legislations.

International treaties may be necessary for such objectives as the setting up of marine reserves. Another international agreement, the Convention on the Conservation of Migratory Species of Wild Animals, protects animals that migrate across the globe and need protection in more than one country. However, the protection of habitats needs to take into account the needs of the local residents for food, fuel and other resources. Even where legislation protects the environment, a lack of enforcement often prevents effective protection. Faced with food shortage, a farmer is likely to plough up a level patch of ground despite it being the last suitable habitat for an endangered species such as the San Quintin kangaroo rat, and even kill the animal as a pest. In this regard, it is desirable to educate the community on the uniqueness of their flora and fauna and the benefits of ecotourism.

Monotypic Habitat

A monotypic habitat is one in which a single species of animal or plant is so dominant as to virtually exclude all other species. An example would be sugarcane; this is planted, burnt and harvested, with herbicides killing weeds and pesticides controlling invertebrates. The monotypic habitat occurs in botanical and zoological contexts, and is a component of conservation biology. In restoration ecology of native plant communities or habitats, some invasive species create monotypic stands that replace and/or prevent other species, especially indigenous ones, from growing there. A dominant colonization can occur from retardant chemicals exuded, nutrient monopolization, or from lack of natural controls such as herbivores or climate, that keep them in balance with their native habitats. The yellow starthistle, *Centaurea solstitialis*, is a botanical monotypic-habitat example of this, currently dominating over 15,000,000 acres (61,000 km^2) in California alone. The non-native freshwater zebra mussel, *Dreissena polymorpha*, that colonizes areas of the Great Lakes and the Mississippi River watershed, is a zoological monotypic-habitat example; the predators that control it in its home-range in Russia are absent and it proliferates abundantly. Even though its name may seem to imply simplicity as compared with polytypic habitats, the monotypic habitat can be complex. Aquatic habitats, such as exotic *Hydrilla* beds, support a similarly rich fauna of macroinvertebrates to a more varied habitat, but the creatures present may differ between the two, affecting small fish and other animals higher up the food chain.

INVASIVE SPECIES

An invasive species is a plant, fungus, or animal species that is not native to a specific location (an introduced species), and which has a tendency to spread to a degree believed to cause damage to the environment, human economy or human health.

One study pointed out widely divergent perceptions of the criteria for invasive species among researchers (p. 135) and concerns with the subjectivity of the term "invasive" (p. 136). Some of the alternate usages of the term are below:

- The term as most often used applies to introduced species (also called "non-indigenous" or "non-native") that adversely affect the habitats and bioregions they invade economically, environmentally, or ecologically. Such invasive species may be either plants or animals and may disrupt by dominating a region, wilderness areas, particular habitats, or wildland–urban interface land from loss of natural controls (such as predators or herbivores). This includes non-native invasive plant species labeled as **exotic pest plants** and **invasive exotics** growing in native plant communities. It has been used in this sense by government organizations as well as conservation groups such as the International Union for Conservation of Nature (IUCN) and the California Native Plant Society. The European Union defines "Invasive Alien Species" as those that are, firstly, outside their natural distribution area, and secondly, threaten biological diversity. It is also used by land managers, botanists, researchers, horticulturalists, conservationists, and the public for noxious weeds. The kudzu vine (*Pueraria lobata*), Andean Pampas grass (*Cortaderia jubata*), and yellow starthistle (*Centaurea solstitialis*) are examples.

- An alternate usage broadens the term to include indigenous or "native" species along with *non-native* species, that have colonized natural areas (p. 136). Deer are an example, considered to be overpopulating their native zones and adjacent suburban gardens, by some in the Northeastern and Pacific Coast regions of the United States.
- Sometimes the term is used to describe a non-native or introduced species that has become widespread (p. 136). However, not every introduced species has adverse effects on the environment. A nonadverse example is the common goldfish (*Carassius auratus*), which is found throughout the United States, but rarely achieves high densities (p. 136).

Beavers from North America constitute an invasive species in Tierra del Fuego, where they have a substantial impact on landscape and local ecology through their dams.

Kudzu, a Japanese vine species invasive in the southeast United States, growing in Atlanta, Georgia.

Causes

Scientists include species- and ecosystem factors among the mechanisms that when combined, establish invasiveness in a newly introduced species.

Species-Based Mechanisms

While all species compete to survive, invasive species appear to have specific traits or specific combinations of traits that allow them to outcompete native species. In some cases, the competition is about rates of growth and reproduction. In other cases, species interact with each other more directly.

Researchers disagree about the usefulness of traits as invasiveness markers. One study found that of a list of invasive and noninvasive species, 86% of the invasive species could be identified from the traits alone. Another study found invasive species tended to have only a small subset of the presumed traits and that many similar traits were found in noninvasive species, requiring other explanations. Common invasive species traits include the following:

- Fast growth
- Rapid reproduction
- High dispersal ability
- Phenotypic plasticity (the ability to alter growth form to suit current conditions)

- Tolerance of a wide range of environmental conditions (Ecological competence)
- Ability to live off of a wide range of food types (generalist)
- Association with humans
- Prior successful invasions

Typically, an introduced species must survive at low population densities before it becomes invasive in a new location. At low population densities, it can be difficult for the introduced species to reproduce and maintain itself in a new location, so a species might reach a location multiple times before it becomes established. Repeated patterns of human movement, such as ships sailing to and from ports or cars driving up and down highways offer repeated opportunities for establishment (also known as a high propagule pressure).

An introduced species might become invasive if it can outcompete native species for resources such as nutrients, light, physical space, water, or food. If these species evolved under great competition or predation, then the new environment may host fewer able competitors, allowing the invader to proliferate quickly. Ecosystems in which are being used to their fullest capacity by native species can be modeled as zero-sum systems in which any gain for the invader is a loss for the native. However, such unilateral competitive superiority (and extinction of native species with increased populations of the invader) is not the rule. Invasive species often coexist with native species for an extended time, and gradually, the superior competitive ability of an invasive species becomes apparent as its population grows larger and denser and it adapts to its new location.

Lantana growing in abandoned citrus plantation; Moshav Sdei Hemed, Israel.

An invasive species might be able to use resources that were previously unavailable to native species, such as deep water sources accessed by a long taproot, or an ability to live on previously uninhabited soil types. For example, barbed goatgrass (*Aegilops triuncialis*) was introduced to California on serpentine soils, which have low water-retention, low nutrient levels, a high magnesium/calcium ratio, and possible heavy metal toxicity. Plant populations on these soils tend to show low density, but goatgrass can form dense stands on these soils and crowd out native species that have adapted poorly to serpentine soils.

Invasive species might alters its environment by releasing chemical compounds, modifying abiotic factors, or affecting the behaviour of herbivores, creating a positive or negative impact on other species. Some species, like *Kalanchoe daigremontana*, produce allelopathic compounds, that might have an inhibitory effect on competing species. Other species like *Stapelia gigantea* facilitates the recruitment of seedlings of other species in arid environments by providing appropriate microclimatic conditions and preventing herbivory in early stages of development.

Another examples are *Centaurea solstitialis* (yellow starthistle) and *Centaurea diffusa* (diffuse knapweed). These Eastern European noxious weeds have spread through the western and West Coast states. Experiments show that 8-hydroxyquinoline, a chemical produced at the root of *C. diffusa*, has a negative effect only on plants that have not co-evolved with it. Such co-evolved native plants have also evolved defenses. *C. diffusa* and *C. solstitialis* do not appear in their native habitats to be overwhelmingly successful competitors. Success or lack of success in one habitat does not necessarily imply success in others. Conversely, examining habitats in which a species is less successful can reveal novel weapons to defeat invasiveness.

Changes in fire regimens are another form of facilitation. *Bromus tectorum*, originally from Eurasia, is highly fire-adapted. It not only spreads rapidly after burning but also increases the frequency and intensity (heat) of fires by providing large amounts of dry detritus during the fire season in western North America. In areas where it is widespread, it has altered the local fire regimen so much that native plants cannot survive the frequent fires, allowing *B. tectorum* to further extend and maintain dominance in its introduced range.

Facilitation also occurs where one species physically modifies a habitat in ways that are advantageous to other species. For example, zebra mussels increase habitat complexity on lake floors, providing crevices in which invertebrates live. This increase in complexity, together with the nutrition provided by the waste products of mussel filter-feeding, increases the density and diversity of benthic invertebrate communities.

Ecosystem-Based Mechanisms

In ecosystems, the amount of available resources and the extent to which those resources are used by organisms determines the effects of additional species on the ecosystem. In stable ecosystems, equilibrium exists in the use of available resources. These mechanisms describe a situation in which the ecosystem has suffered a disturbance, which changes the fundamental nature of the ecosystem.

When changes such as a forest fire occur, normal succession favors native grasses and forbs. An introduced species that can spread faster than natives can use resources that would have been available to native species, squeezing them out. Nitrogen and phosphorus are often the limiting factors in these situations.

Every species occupies a *niche* in its native ecosystem; some species fill large and varied roles, while others are highly specialized. Some invading species fill niches that are not used by native species, and they also can create new niches. An example of this type can be found within the *Lampropholis delicata* species of skink.

Ecosystem changes can alter specie's distributions. For example, edge effects describe what happens when part of an ecosystem is disturbed as when land is cleared for agriculture. The boundary between remaining undisturbed habitat and the newly cleared land itself forms a distinct habitat, creating new winners and losers and possibly hosting species that would not thrive outside the boundary habitat.

One interesting finding in studies of invasive species has shown that introduced populations have great potential for rapid adaptation and this is used to explain how so many introduced species are able to establish and become invasive in new environments. When bottlenecks and founder effects cause a great decrease in the population size and may constrict genetic variation, the individuals begin to show additive variance as opposed to epistatic variance. This conversion can actually lead to increased variance in the founding populations which then allows for rapid adaptive evolution. Following invasion events, selection may initially act on the capacity to disperse as well as physiological tolerance to the new stressors in the environment. Adaptation then proceeds to respond to the selective pressures of the new environment. These responses would most likely be due to temperature and climate change, or the presence of native species whether it be predator or prey. Adaptations include changes in morphology, physiology, phenology, and plasticity.

Rapid adaptive evolution in these species leads to offspring that have higher fitness and are better suited for their environment. Intraspecific phenotypic plasticity, pre-adaptation and post-introduction evolution are all major factors in adaptive evolution. Plasticity in populations allows room for changes to better suit the individual in its environment. This is key in adaptive evolution because the main goal is how to best be suited to the ecosystem that the species has been introduced. The ability to accomplish this as quickly as possible will lead to a population with a very high fitness. Pre-adaptations and evolution after the initial introduction also play a role in the success of the introduced species. If the species has adapted to a similar ecosystem or contains traits that happen to be well suited to the area that it is introduced, it is more likely to fare better in the new environment. This, in addition to evolution that takes place after introduction, all determine if the species will be able to become established in the new ecosystem and if it will reproduce and thrive.

Ecology

Traits of Invaded Ecosystems

In 1958, Charles S. Elton claimed that ecosystems with higher species diversity were less subject to invasive species because of fewer available niches. Other ecologists later pointed to highly diverse, but heavily invaded ecosystems and argued that ecosystems with high species diversity were more susceptible to invasion.

This debate hinged on the spatial scale at which invasion studies were performed, and the issue of how diversity affects susceptibility remained unresolved as of 2011. Small-scale studies tended to show a negative relationship between diversity and invasion, while large-scale studies tended to show the reverse. The latter result may be a side-effect of invasives' ability to capitalize on increased resource availability and weaker species interactions that are more common when larger samples are considered.

The brown tree snake (*Boiga irregularis*)

Invasion was more likely in ecosystems that were similar to the one in which the potential invader evolved. Island ecosystems may be more prone to invasion because their species faced few strong competitors and predators, or because their distance from colonizing species populations makes them more likely to have "open" niches. An example of this phenomenon was the decimation of native bird populations on Guam by the invasive brown tree snake. Conversely, invaded ecosystems may lack the natural competitors and predators that check invasives' growth in their native ecosystems.

Invaded ecosystems may have experienced disturbance, typically human-induced. Such a disturbance may give invasive species a chance to establish themselves with less competition from natives less able to adapt to a disturbed ecosystem.

Vectors

Non-native species have many *vectors*, including biogenic vectors, but most invasions are associated with human activity. Natural range extensions are common in many species, but the rate and magnitude of human-mediated extensions in these species tend to be much larger than natural extensions, and humans typically carry specimens greater distances than natural forces. An early human vector occurred when prehistoric humans introduced the Pacific rat (*Rattus exulans*) to Polynesia.

Chinese mitten crab (*Eriocheir sinensis*)

Vectors include plants or seeds imported for horticulture. The pet trade moves animals across borders, where they can escape and become invasive. Organisms stow away on transport vehicles. Ballast water taken up at sea and released in port by transoceanic vessels is the largest vector for non-native aquatic species invasions. Around the world on the average day, more than 3,000 different species of aquatic life may be transported on these vessels. For example, freshwater zebra mussels, native to the Black, Caspian and Azov seas, probably reached the Great Lakes via ballast water from a transoceanic vessel. Although the zebra mussel invasion was first noted in 1988, and a mitigation plan was successfully implemented shortly thereafter, the plan had (and continued to have as of 2005) a serious flaw or loophole, whereby ships that are loaded with cargo when they reach the Seaway need not be tested, but all the same they transfer ballast 'puddles' between Seaway ports.

The arrival of invasive propagules to a new site is a function of the site's invasibility.

Species have also been introduced intentionally. For example, to feel more "at home," American colonists formed "Acclimation Societies" that repeatedly imported birds that were native to Europe to North America and other distant lands. In 2008, U.S. postal workers in Pennsylvania noticed noises coming from inside a box from Taiwan; the box contained more than two dozen live beetles. Agricultural Research Service entomologists identified them as rhinoceros beetle, hercules beetle, and king stag beetle. Because these species were not native to the U.S., they could have threatened native ecosystems. To prevent exotic species from becoming a problem in the U.S., special handling and permits are required when living materials are shipped from foreign countries. USDA programs such as Smuggling Interdiction and Trade Compliance (SITC) attempt to prevent exotic species outbreaks in America.

Economics plays a major role in exotic species introduction. High demand for the valuable Chinese mitten crab is one explanation for the possible intentional release of the species in foreign waters.

Impacts of Wildfire

Invasive species often exploit disturbances to an ecosystem (wildfires, roads, foot trails) to colonize an area. Large wildfires can sterilize soils, while adding a variety of nutrients. In the resulting free-for-all, formerly entrenched species lose their advantage, leaving more room for invasives. In such circumstances plants that can regenerate from their roots have an advantage. Non-natives with this ability can benefit from a low intensity fire burns that removes surface vegetation, leaving natives that rely on seeds for propagation to find their niches occupied when their seeds finally sprout.

Impact of Wildfire Suppression on Spreading

Wildfires often occur in remote areas, needing fire suppression crews to travel through pristine forest to reach the site. The crews can bring invasive seeds with them. If any of these stowaway seeds become established, a thriving colony of invasives can erupt in as few as six weeks, after which controlling the outbreak can need years of continued attention to prevent further spread. Also, disturbing the soil surface, such as cutting firebreaks, destroys native cover, exposes soil, and can accelerate invasions. In suburban and wildland-urban interface areas, the vegetation clearance and brush removal ordinances of municipalities for defensible space can result in excessive removal of native shrubs and perennials that exposes the soil to more light and less competition for invasive plant species.

Fire suppression vehicles are often major culprits in such outbreaks, as the vehicles are often driven on back roads often overgrown with invasive plant species. The undercarriage of the vehicle becomes a prime vessel of transport. In response, on large fires, washing stations "decontaminate" vehicles before engaging in suppression activities. Large wildfires attract firefighters from remote places, further increasing the potential for seed transport.

Effects

An American alligator attacking a Burmese python in Florida; the Burmese python is an invasive species which is posing a threat to many indigenous species, including the alligator

Ecological

Land clearing and human habitation put significant pressure on local species. Disturbed habitats are prone to invasions that can have adverse effects on local ecosystems, changing ecosystem functions. A species of wetland plant known as 'ae'ae in Hawaii (the indigenous *Bacopa monnieri*) is regarded as a pest species in artificially manipulated water bird refuges because it quickly covers shallow mudflats established for endangered Hawaiian stilt (*Himantopus mexicanus knudseni*), making these undesirable feeding areas for the birds.

Multiple successive introductions of different non-native species can have interactive effects; the introduction of a second non-native species can enable the first invasive species to flourish. Examples of this are the introductions of the amethyst gem clam (*Gemma gemma*) and the European green crab (*Carcinus maenas*). The gem clam was introduced into California's Bodega Harbor from the East Coast of the United States a century ago. It had been found in small quantities in the harbor but had never displaced the native clam species (*Nutricola* spp.). In the mid-1990s, the introduction of the European green crab, found to prey preferentially on the native clams, resulted in a decline of the native clams and an increase of the introduced clam populations.

In the Waterberg region of South Africa, cattle grazing over the past six centuries has allowed invasive scrub and small trees to displace much of the original grassland, resulting in a massive reduction in forage for native bovids and other grazers. Since the 1970s, large scale efforts have been underway to reduce invasive species; partial success has led to re-establishment of many species that had dwindled or left the region. Examples of these species are giraffe, blue wildebeest, impala, kudu and white rhino.

Invasive species can change the functions of ecosystems. For example, invasive plants can alter the fire regimen (cheatgrass, *Bromus tectorum*), nutrient cycling (smooth cordgrass *Spartina alterniflora*), and hydrology (*Tamarix*) in native ecosystems. Invasive species that are closely related to rare native species have the potential to hybridize with the native species. Harmful effects of hybridization have led to a decline and even extinction of native species. For example, hybridization with introduced cordgrass, *Spartina alterniflora*, threatens the existence of California cordgrass (*Spartina foliosa*) in San Francisco Bay. Invasive species cause competition for native species and because of this 400 of the 958 endangered species under the Endangered Species Act are at risk.

Geomorphological

Primary geomorphological effects of invasive plants are bioconstruction and bioprotection. For example, Kudzu Pueraria montana, a vine native to Asia was widely introduced in the southeastern USA in the early 20th century to control soil erosion. While primary effects of invasive animals are bioturbation, bioerosion, and bioconstruction. For example, invasion of Chinese mitten crab Eriocheir sinensis have resulted to higher bioturbation and bioerosion rates.

Economic

Economic Opportunities

Some invasions offer potential commercial benefits. For instance, silver carp and common carp can be harvested for human food and exported to markets already familiar with the product, or processed into pet foods, or mink feed. Water hyacinth can be turned into fuel by methane digesters, and other invasive plants can also be harvested and utilized as a source of bioenergy.

Benefits

Non-native species can have benefits. Asian oysters, for example, filter water pollutants better than native oysters. They also grow faster and withstand disease better than natives. Biologists are currently considering releasing this mollusk in the Chesapeake Bay to help restore oyster stocks and remove pollution. A recent study by the Johns Hopkins School of Public Health found the Asian oyster could significantly benefit the bay's deteriorating water quality. Additionally, some species have invaded an area so long ago that they have found their own beneficial niche in the environment, a term referred to as naturalisation. For example, *L. leucozonium*, shown by population genetic analysis to be an invasive species in North America, has become an important pollinator of caneberry as well as cucurbit, apple trees, and blueberry bushes.

Invasivorism

Invasive species are flora and fauna whose introduction into a habitat disrupts the native ecosystem. In response, Invasivorism is a movement that explores the idea of eating invasive species in order to control, reduce, or eliminate their populations. Chefs from around the world have begun seeking out and using invasive species as alternative ingredients. Miya's of New Haven, Connecticut created the first invasive species menu in the world. Skeptics point out that once a foreign species has entrenched itself in a new place—such as the Indo-Pacific lionfish that has now virtually taken over the waters of the Western Atlantic, Caribbean and Gulf of Mexico—eradication is almost impossible. Critics argue that encouraging consumption might have the unintended effect of spreading harmful species even more widely.

A dish that features whole fried invasive lionfish at Fish Fish of Miami, Florida.

Proponents of invasivorism argue that humans have the ability to eat away any species that it has an appetite for, pointing to the many animals which humans have been able to hunt to extinction - such as the Dodo bird, the Caribbean monk seal, and the Passenger pigeon. Proponents of invasivorism also point to the success that Jamaica has had in significantly decreasing the population of lionfish by encouraging the consumption of the fish.

Costs

Economic costs from invasive species can be separated into direct costs through production loss in agriculture and forestry, and management costs. Estimated damage and control cost of invasive species in the U.S. alone amount to more than $138 billion annually. Economic losses can also occur through loss of recreational and tourism revenues. When economic costs of invasions are calculated as production loss and management costs, they are low because they do not consider environmental damage; if monetary values were assigned to the extinction of species, loss in biodiversity, and loss of ecosystem services, costs from impacts of invasive species would drastically increase. The following examples from different sectors of the economy demonstrate the impact of biological invasions.

Plant Industry

Weeds reduce yield in agriculture, though they may provide essential nutrients. Some deep-rooted weeds can "mine" nutrients (see dynamic accumulator) from the subsoil and deposit them on the topsoil, while others provide habitat for beneficial insects or provide foods for pest species. Many weed species are accidental introductions that accompany seeds and imported plant material. Many introduced weeds in pastures compete with native forage plants, threaten young cattle (e.g., leafy spurge, *Euphorbia esula*) or are unpalatable because of thorns and spines (e.g., yellow starthistle). Forage loss from invasive weeds on pastures amounts to nearly US$1 billion in the U.S. alone. A decline in pollinator services and loss of fruit production has been caused by honey bees infected by the invasive varroa mite. Introduced rats (*Rattus rattus* and *R. norvegicus*) have become serious pests on farms, destroying stored grains.

Invasive plant pathogens and insect vectors for plant diseases can also suppress agricultural yields and nursery stock. Citrus greening is a bacterial disease vectored by the invasive Asian citrus psyllid (ACP). Because of the impacts of this disease on citrus crops, citrus is under quarantine and highly regulated in areas where ACP has been found.

Aquaculture

Aquaculture is a very common vector of species introductions – mainly of species with economic potential (e.g., Oreochromis niloticus).

Forestry

Poster asking campers to not move firewood around, avoiding the spread of invasive species.

The unintentional introduction of forest pest species and plant pathogens can change forest ecology and damage the timber industry. Overall, forest ecosystems in the U.S. are widely invaded by exotic pests, plants, and pathogens.

The Asian long-horned beetle (*Anoplophora glabripennis*) was first introduced into the U.S. in 1996, and was expected to infect and damage millions of acres of hardwood trees. As of 2005. thirty million dollars had been spent in attempts to eradicate this pest and protect millions of trees in the affected regions. The woolly adelgid has inflicted damage on old-growth spruce, fir and hemlock forests and damages the Christmas tree industry. And the chestnut blight fungus (*Cryphonectria parasitica*) and Dutch elm disease (*Ophiostoma novo-ulmi*) are two plant pathogens with serious impacts on these two species, and forest health. Garlic mustard, *Alliaria petiolata*, is one of the most problematic invasive plant species in eastern North American forests. The characteristics of garlic mustard are slightly different from those of the surrounding native plants, which results in a highly successful species that is altering the composition and function of the native communities it invades. When garlic mustard invades the understory of a forest, it affects the growth rate of tree seedlings, which is likely to alter forest regeneration of impact forest composition in the future.

Tourism and Recreation

Invasive species can impact outdoor recreation, such as fishing, hunting, hiking, wildlife viewing, and water-based activities. They can damage a wide array of environmental services that are important to recreation, including, but not limited to, water quality and quantity, plant and animal diversity, and species abundance. Eiswerth states, "very little research has been performed to estimate the corresponding economic losses at spatial scales such as regions, states, and watersheds." Eurasian watermilfoil (*Myriophyllum spicatum*) in parts of the US, fill lakes with plants complicating fishing and boating. The very loud call of the introduced common coqui depresses real estate values in affected neighborhoods of Hawaii.

Health

Encroachment of humans into previously remote ecosystems has exposed exotic diseases such as HIV to the wider population. Introduced birds (e.g. pigeons), rodents and insects (e.g. mosquito, flea, louse and tsetse fly pests) can serve as vectors and reservoirs of human afflictions. The introduced Chinese mitten crabs are carriers of Asian lung fluke. Throughout recorded history, epidemics of human diseases, such as malaria, yellow fever, typhus, and bubonic plague, spread via these vectors. A recent example of an introduced disease is the spread of the West Nile virus, which killed humans, birds, mammals, and reptiles. Waterborne disease agents, such as cholera bacteria (*Vibrio cholerae*), and causative agents of harmful algal blooms are often transported via ballast water. Invasive species and accompanying control efforts can have long term public health implications. For instance, pesticides applied to treat a particular pest species could pollute soil and surface water.

Biodiversity

Biotic invasion is considered one of the five top drivers for global biodiversity loss and is increasing because of tourism and globalization. This may be particularly true in inadequately regulated fresh water systems, though quarantines and ballast water rules have improved the situation.

Invasive species may drive local native species to extinction via competitive exclusion, niche displacement, or hybridisation with related native species. Therefore, besides their economic ramifications, alien invasions may result in extensive changes in the structure, composition and global distribution of the biota of sites of introduction, leading ultimately to the homogenisation of the world's fauna and flora and the loss of biodiversity. Nevertheless, it is difficult to unequivocally attribute extinctions to a species invasion, and the few scientific studies that have done so have been with animal taxa. Concern over the impacts of invasive species on biodiversity must therefore consider the actual evidence (either ecological or economic), in relation to the potential risk.

Alien invasive species *Parthenium hysterophorus* smothering native flora in Achanakmar Tiger Reserve, Bilaspur, Chhattisgarh, India.

Genetic Pollution

Native species can be threatened with extinction through the process of *genetic pollution*. Genetic pollution is unintentional hybridization and introgression, which leads to homogenization or replacement of local genotypes as a result of either a numerical or fitness advantage of the introduced species. Genetic pollution can operate either through introduction or through habitat modification, bringing previously isolated species into contact. Hybrids resulting from rare species that interbreed with abundant species can swamp the rarer species' gene pool. This is not always apparent from morphological observations alone. Some degree of gene flow is normal, and preserves constellations of genes and genotypes. An example of this is the interbreeding of migrating coyotes with the red wolf, in areas of eastern North Carolina where the red wolf was reintroduced.

Geography

Satellite photograph of deforestation in Bolivia. Originally dry tropical forest, the land is being cleared for soybean cultivation.

Biodiversity hotspots are chiefly tropical regions that feature high concentrations of endemic species and, when all hotspots are combined, may contain over half of the world's terrestrial species. These hotspots are suffering from habitat loss and destruction. Most of the natural habitat on islands and in areas of high human population density has already been destroyed (WRI, 2003). Islands suffering extreme habitat destruction include New Zealand, Madagascar, the Philippines, and Japan. South and east Asia — especially China, India, Malaysia, Indonesia, and Japan — and many areas in West Africa have extremely dense human populations that allow little room for natural habitat. Marine areas close to highly populated coastal cities also face degradation of their coral reefs or other marine habitat. These areas include the eastern coasts of Asia and Africa, northern coasts of South America, and the Caribbean Sea and its associated islands.

Regions of unsustainable agriculture or unstable governments, which may go hand-in-hand, typically experience high rates of habitat destruction. Central America, Sub-Saharan Africa, and the Amazonian tropical rainforest areas of South America are the main regions with unsustainable agricultural practices or government mismanagement.

Areas of high agricultural output tend to have the highest extent of habitat destruction. In the U.S., less than 25% of native vegetation remains in many parts of the East and Midwest. Only 15% of land area remains unmodified by human activities in all of Europe.

Ecosystems

Jungle burned for agriculture in southern Mexico.

Tropical rainforests have received most of the attention concerning the destruction of habitat. From the approximately 16 million square kilometers of tropical rainforest habitat that originally existed worldwide, less than 9 million square kilometers remain today. The current rate of deforestation is 160,000 square kilometers per year, which equates to a loss of approximately 1% of original forest habitat each year.

Other forest ecosystems have suffered as much or more destruction as tropical rainforests. Farming and logging have severely disturbed at least 94% of temperate broadleaf forests; many old growth forest stands have lost more than 98% of their previous area because of human activities. Tropical deciduous dry forests are easier to clear and burn and are more suitable for agriculture and cattle ranching than tropical rainforests; consequently, less than 0.1% of dry forests in Central America's Pacific Coast and less than 8% in Madagascar remain from their original extents.

Plains and desert areas have been degraded to a lesser extent. Only 10-20% of the world's drylands, which include temperate grasslands, savannas, and shrublands, scrub, and deciduous forests, have been somewhat degraded. But included in that 10-20% of land is the approximately 9 million square kilometers of seasonally dry-lands that humans have converted to deserts through the process of desertification. The tallgrass prairies of North America, on the other hand, have less than 3% of natural habitat remaining that has not been converted to farmland.

Wetlands and marine areas have endured high levels of habitat destruction. More than 50% of wetlands in the U.S. have been destroyed in just the last 200 years. Between 60% and 70% of European wetlands have been completely destroyed. About one-fifth (20%) of marine coastal areas have been highly modified by humans. One-fifth of coral reefs have also been destroyed, and another fifth has been severely degraded by overfishing, pollution, and invasive species; 90% of the Philippines' coral reefs alone have been destroyed. Finally, over 35% mangrove ecosystems worldwide have been destroyed.

Farmers near newly cleared land within Taman Nasional Kerinci Seblat (Kerinci Seblat National Park), Sumatra.

Natural Causes

Habitat destruction through natural processes such as volcanism, fire, and climate change is well documented in the fossil record. One study shows that habitat fragmentation of tropical rainforests in Euramerica 300 million years ago led to a great loss of amphibian diversity, but simultaneously the drier climate spurred on a burst of diversity among reptiles.

Human Causes

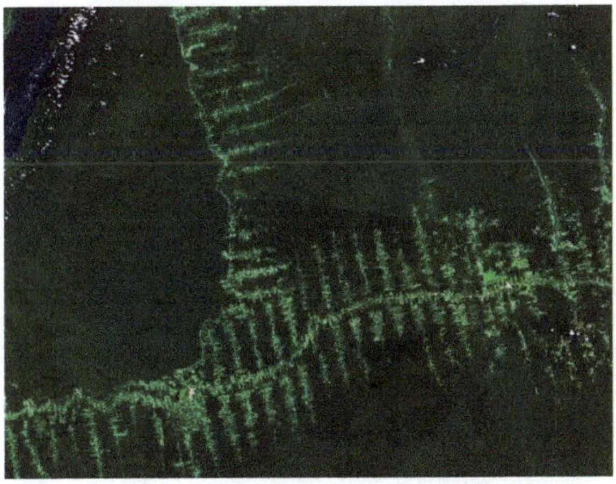

Deforestation and roads in Amazonia, the Amazon Rainforest.

Habitat destruction caused by humans includes land conversion from forests, etc. to arable land, urban sprawl, infrastructure development, and other anthropogenic changes to the characteristics of land. Habitat degradation, fragmentation, and pollution are aspects of habitat destruction caused by humans that do not necessarily involve over destruction of habitat, yet result in habitat collapse. Desertification, deforestation, and coral reef degradation are specific types of habitat destruction for those areas (deserts, forests, coral reefs).

Geist and Lambin (2002) assessed 152 case studies of net losses of tropical forest cover to determine any patterns in the proximate and underlying causes of tropical deforestation. Their results, yielded as percentages of the case studies in which each parameter was a significant factor, provide a quantitative prioritization of which proximate and underlying causes were the most significant. The proximate causes were clustered into broad categories of agricultural expansion (96%), infrastructure expansion (72%), and wood extraction (67%). Therefore, according to this study, forest conversion to agriculture is the main land use change responsible for tropical deforestation. The specific categories reveal further insight into the specific causes of tropical deforestation: transport extension (64%), commercial wood extraction (52%), permanent cultivation (48%), cattle ranching (46%), shifting (slash and burn) cultivation (41%), subsistence agriculture (40%), and fuel wood extraction for domestic use (28%). One result is that shifting cultivation is not the primary cause of deforestation in all world regions, while transport extension (including the construction of new roads) is the largest single proximate factor responsible for deforestation.

Drivers

Nanjing Road in Shanghai

While the above-mentioned activities are the proximal or direct causes of habitat destruction in that they actually destroy habitat, this still does not identify why humans destroy habitat. The forces that cause humans to destroy habitat are known as *drivers* of habitat destruction. Demographic, economic, sociopolitical, scientific and technological, and cultural drivers all contribute to habitat destruction.

Demographic drivers include the expanding human population; rate of population increase over time; spatial distribution of people in a given area (urban versus rural), ecosystem type, and country; and the combined effects of poverty, age, family planning, gender, and education status of people in certain areas. Most of the exponential human population growth worldwide is occurring in or close to biodiversity hotspots. This may explain why human population density accounts for 87.9% of the variation in numbers of threatened species across 114 countries, providing indisputable evidence that people play the largest role in decreasing biodiversity. The boom in human population and migration of people into such species-rich regions are making conservation efforts not only more urgent but also more likely to conflict with local human interests. The high local population density in such areas is directly correlated to the poverty status of the local people, most of whom lacking an education and family planning.

From the Geist and Lambin (2002) study described in the previous section, the underlying driving forces were prioritized as follows (with the percent of the 152 cases the factor played a significant role in): economic factors (81%), institutional or policy factors (78%), technological factors (70%), cultural or socio-political factors (66%), and demographic factors (61%). The main economic factors included commercialization and growth of timber markets (68%), which are driven by national and international demands; urban industrial growth (38%); low domestic costs for land, labor, fuel, and timber (32%); and increases in product prices mainly for cash crops (25%). Institutional and policy factors included formal pro-deforestation policies on land development (40%), economic growth including colonization and infrastructure improvement (34%), and subsidies for land-based activities (26%); property rights and land-tenure insecurity (44%); and policy failures such as corruption, lawlessness, or mismanagement (42%). The main technological factor was the poor application of technology in the wood industry (45%), which leads to wasteful logging practices. Within the broad category of cultural and sociopolitical factors are public attitudes and values (63%), individual/household behavior (53%), public unconcern toward forest environments (43%), missing basic values (36%), and unconcern by individuals (32%). Demographic factors were the in-migration of colonizing settlers into sparsely populated forest areas (38%) and growing population density — a result of the first factor — in those areas (25%).

There are also feedbacks and interactions among the proximate and underlying causes of deforestation that can amplify the process. Road construction has the largest feedback effect, because it interacts with—and leads to—the establishment of new settlements and more people, which causes a growth in wood (logging) and food markets. Growth in these markets, in turn, progresses the commercialization of agriculture and logging industries. When these industries become commercialized, they must become more efficient by utilizing larger or more modern machinery that often are worse on the habitat than traditional farming and logging methods. Either way, more land is cleared more rapidly for commercial markets. This common feedback example manifests just how closely related the proximate and underlying causes are to each other.

Impact on Human Population

The draining and development of coastal wetlands that previously protected the Gulf Coast contributed to severe flooding in New Orleans, Louisiana in the aftermath of Hurricane Katrina.

Habitat destruction vastly increases an area's vulnerability to natural disasters like flood and drought, crop failure, spread of disease, and water contamination. On the other hand, a healthy ecosystem with good management practices will reduce the chance of these events happening, or will at least mitigate adverse impacts.

Agricultural land can actually suffer from the destruction of the surrounding landscape. Over the past 50 years, the destruction of habitat surrounding agricultural land has degraded approximately 40% of agricultural land worldwide via erosion, salinization, compaction, nutrient depletion, pollution, and urbanization. Humans also lose direct uses of natural habitat when habitat is destroyed. Aesthetic uses such as birdwatching, recreational uses like hunting and fishing, and ecotourism usually rely upon virtually undisturbed habitat. Many people value the complexity of the natural world and are disturbed by the loss of natural habitats and animal or plant species worldwide.

Probably the most profound impact that habitat destruction has on people is the loss of many valuable ecosystem services. Habitat destruction has altered nitrogen, phosphorus, sulfur, and carbon cycles, which has increased the frequency and severity of acid rain, algal blooms, and fish kills in rivers and oceans and contributed tremendously to global climate change. One ecosystem service whose significance is becoming more realized is climate regulation. On a local scale, trees provide windbreaks and shade; on a regional scale, plant transpiration recycles rainwater and maintains constant annual rainfall; on a global scale, plants (especially trees from tropical rainforests) from around the world counter the accumulation of greenhouse gases in the atmosphere by sequestering carbon dioxide through photosynthesis. Other ecosystem services that are diminished or lost altogether as a result of habitat destruction include watershed management, nitrogen fixation, oxygen production, pollination, waste treatment (i.e., the breaking down and immobilization of toxic pollutants), and nutrient recycling of sewage or agricultural runoff.

The loss of trees from the tropical rainforests alone represents a substantial diminishing of the earth's ability to produce oxygen and use up carbon dioxide. These services are becoming even more important as increasing carbon dioxide levels is one of the main contributors to global climate change.

The loss of biodiversity may not directly affect humans, but the indirect effects of losing many species as well as the diversity of ecosystems in general are enormous. When biodiversity is lost, the environment loses many species that provide valuable and unique roles to the ecosystem. The environment and all its inhabitants rely on biodiversity to recover from extreme environmental conditions. When too much biodiversity is lost, a catastrophic event such as an earthquake, flood, or volcanic eruption could cause an ecosystem to crash, and humans would obviously suffer from that. Loss of biodiversity also means that humans are losing animals that could have served as biological control agents and plants that could potentially provide higher-yielding crop varieties, pharmaceutical drugs to cure existing or future diseases or cancer, and new resistant crop varieties for agricultural species susceptible to pesticide-resistant insects or virulent strains of fungi, viruses, and bacteria.

The negative effects of habitat destruction usually impact rural populations more directly than urban populations. Across the globe, poor people suffer the most when natural habitat is destroyed, because less natural habitat means less natural resources per capita, yet wealthier people and countries simply have to pay more to continue to receive more than their per capita share of natural resources.

Another way to view the negative effects of habitat destruction is to look at the opportunity cost of keeping an area undisturbed. In other words, what are people losing out on by taking away a given habitat? A country may increase its food supply by converting forest land to row-crop agriculture, but the value of the same land may be much larger when it can supply natural resources or services such as clean water, timber, ecotourism, or flood regulation and drought control.

Outlook

The rapid expansion of the global human population is increasing the world's food requirement substantially. Simple logic instructs that more people will require more food.

In fact, as the world's population increases dramatically, agricultural output will need to increase by at least 50%, over the next 30 years. In the past, continually moving to new land and soils provided a boost in food production to appease the global food demand. That easy fix will no longer be available, however, as more than 98% of all land suitable for agriculture is already in use or degraded beyond repair.

The impending global food crisis will be a major source of habitat destruction. Commercial farmers are going to become desperate to produce more food from the same amount of land, so they will use more fertilizers and less concern for the environment to meet the market demand. Others will seek out new land or will convert other land-uses to agriculture. Agricultural intensification will become widespread at the cost of the environment and its inhabitants. Species will be pushed out of their habitat either directly by habitat destruction or indirectly by fragmentation, degradation, or pollution. Any efforts to protect the world's remaining natural habitat and biodiversity will compete directly with humans' growing demand for natural resources, especially new agricultural lands.

Solutions

Chelonia mydas on a Hawaiian coral reef. Although the endangered species is protected, habitat loss from human development is a major reason for the loss of green turtle nesting beaches.

In most cases of tropical deforestation, three to four underlying causes are driving two to three proximate causes. This means that a universal policy for controlling tropical deforestation would not be able to address the unique combination of proximate and underlying causes of deforestation in each country. Before any local, national, or international deforestation policies are written and enforced, governmental leaders must acquire a detailed understanding of the complex combination of proximate causes and underlying driving forces of deforestation in a given area or country. This concept, along with many other results about tropical deforestation from the Geist and Lambin study, can easily be applied to habitat destruction in general. Governmental leaders need to take action by addressing the underlying driving forces, rather than merely regulating the proximate causes. In a broader sense, governmental bodies at a local, national, and international scale need to emphasize the following:

- Considering the many irreplaceable ecosystem services provided by natural habitats.
- Protecting remaining intact sections of natural habitat.
- Educating the public about the importance of natural habitat and biodiversity.
- Developing family planning programs in areas of rapid population growth.
- Finding ecological ways to increase agricultural output without increasing the total land in production.
- Preserving habitat corridors to minimize prior damage from fragmented habitats.
- Reduce human population and expansion.

REFERENCES

1. Barbault, R. and S. D. Sastrapradja. 1995. Generation, maintenance and loss of biodiversity. Global Biodiversity Assessment, Cambridge Univ. Press, Cambridge pp. 193–274.
2. Burke, L., Y. Kura, K. Kassem, C. Ravenga, M. Spalding, and D. McAllister. 2000. Pilot Assessment of Global Ecosystems: Coastal Ecosystems. World Resources Institute, Washington, D.C.
3. Cincotta, R.P., and R. Engelman. 2000. Nature's place: human population density and the future of biological diversity. Population Action International. Washington, D.C.
4. Geist H. J.; Lambin E. E. (2002). "Proximate causes and underlying driving forces of tropical deforestation". BioScience. 52 (2): 143–150. doi:10.1641/0006-3568(2002)052[0143:PCAUDF]2.0.CO;2.
5. Kauffman, J. B. and D. A. Pyke. 2001. Range ecology, global livestock influences. In S. A. Levin (ed.), Encyclopedia of Biodiversity 5: 33-52. Academic Press, San Diego, CA.
6. Laurance W. F. (1999). "Reflections on the tropical deforestation crisis". Biological Conservation. 91: 109–117. doi:10.1016/S0006-3207(99)00088-9.
7. McKee J. K.; Sciulli P.W.; Fooce C. D.; Waite T. A. (2003). "Forecasting global biodiversity threats associated with human population growth". Biological Conservation. 115: 161–164. doi:10.1016/s0006-3207(03)00099-5.
8. Millennium Ecosystem Assessment (Program). 2005. Ecosystems and Human Well-Being. Millennium Ecosystem Assessment. Island Press, Covelo, CA.

9. Primack, R. B. 2006. Essentials of Conservation Biology. 4th Ed. Habitat destruction, pages 177-188. Sinauer Associates, Sunderland, MA.
10. Pimm Stuart L.; Raven Peter (2000). "Biodiversity: Extinction by numbers". Nature. 403 (6772): 843–845. doi:10.1038/35002708. PMID 10706267.
11. Ravenga, C., J. Brunner, N. Henninger, K. Kassem, and R. Payne. 2000. Pilot Analysis of Global Ecosystems: Wetland Ecosystems. World Resources Institute, Washington, D.C.
12. Sahney S.; Benton M.J.; Falcon-Lang H.J. (2010). "Rainforest collapse triggered Pennsylvanian tetrapod diversification in Euramerica". Geology. 38: 1079–1082. doi:10.1130/G31182.1.
13. Sanderson E. W.; Jaiteh M.; Levy M. A.; Redford K. H.; Wannebo A. V.; Woolmer G. (2002). "The human footprint and the last of the wild". BioScience. 52 (10): 891–904. doi:10.1641/0006-3568(2002)052[0891:thfatl]2.0.co;2.
14. Scholes, R. J. and R. Biggs (eds.). 2004. Ecosystem services in Southern Africa: a regional assessment. The regional scale component of the Southern African Millennium Ecosystem Assessment. CSIR, Pretoria, South Africa.
15. Stein, B. A., L. S. Kutner, and J. S. Adams (eds.). 2000. Precious Heritage: The Status of Biodiversity in the United States. Oxford University Press, New York.
16. Temple S. A. (1986). "The problem of avian extinctions". Ornithology. 3: 453–485. doi:10.1007/978-1-4615-6784-4_11.
17. Tibbetts John (2006). "Louisiana's Wetlands: A Lesson in Nature Appreciation". Environ Health Perspect. 114 (1): A40–A43. doi:10.1289/ehp.114-a40. PMC 1332684 . PMID 16393646.
18. Tilman D.; Fargione J.; Wolff B.; D'Antonio C.; Dobson A.; Howarth R.; Schindler D.; Schlesinger W. H.; Simberloff D.; et al. (2001). "Forecasting agriculturally driven global environmental change". Science. 292: 281–284. doi:10.1126/science.1057544. PMID 11303102.
19. White, R. P., S. Murray, and M. Rohweder. 2000. Pilot Assessment of Global Ecosystems: Grassland Ecosystems. World Resources Institute, Washington, D. C.
20. WRI. 2003. World Resources 2002-2004: Decisions for the Earth: Balance, voice, and power. 328 pp. World Resources Institute, Washington, D.C.
21. "Habitat". Dictionary.com. Retrieved 16 May 2016.
22. "Habitat". Merriam-Webster Dictionary. Retrieved 4 June 2016.
23. "Biotope". Oxford Dictionaries. Retrieved 4 June 2016.
24. Everyman's Encyclopedia; Volume 4. J.M. Dent. 1967. p. 581. ASIN B0015GRC04.
25. Richards, O.W. (1940). "The biology of the small white butterfly (Pieris rapae), with special reference to the factors controlling its abundance". Journal of Animal Ecology. 9 (2): 243–288. doi:10.2307/1459.
26. Spitzer, L.; Benes, J.; Dandova,J.; Jaskova,V.; Konvicka, M. (2009). "The Large Blue butterfly ("Phengaris [Maculinea] arion"), as a conservation umbrella on

a landscape scale: The case of the Czech Carpathians". Ecological Indicators. 9: 1056–1063. doi:10.1016/j.ecolind.2008.12.006.

27. Sutherland, William J.; Hill, David A. (1995). Managing Habitats for Conservation. Cambridge University Press. p. 6. ISBN 978-0-521-44776-8.
28. Richard J. Huggett (2004). Fundamentals of Biogeography. Psychology Press. p. 146. ISBN 978-0-415-32347-5.
29. "Invasive species". National Wildlife Federation. Retrieved 24 May 2016.
30. Breed, Michael D.; Moore, Janice (2011). Animal Behavior. Academic Press. p. 248. ISBN 978-0-08-091992-8.
31. "Habitats". BBC Nature. Retrieved 24 May 2016.
32. Cook, C.D.K.; Gut, B.J.; Rix, E.M.; Schneller, J. (1974). Water Plants of the World: A Manual for the Identification of the Genera of Freshwater Macrophytes. Springer Science & Business Media. p. 7. ISBN 978-90-6193-024-2.
33. Roff, John (2013). Marine Conservation Ecology. Routledge. p. 105. ISBN 978-1-136-53838-4.
34. "Couch's spadefoot (Scaphiopus couchi)". Arizona–Sonora Desert Museum. Retrieved 16 May 2016.
35. Witham, Carol W. (1998). Ecology, Conservation, and Management of Vernal Pool Ecosystems. California Native Plant Society. p. 1. ISBN 978-0-943460-37-6.
36. Green, Scott. "Fairy shrimp". The Vernal Pool Association. Retrieved 17 May 2016.
37. Walker, Matt (21 May 2015). "The most extreme fish on Earth". BBC Earth. Retrieved 17 May 2016.
38. Read, Nicholas (2012). City Critters: Wildlife in the Urban Jungle. Orca Book Publishers. p. 2. ISBN 978-1-55469-394-8.
39. John G. Kelcey, John G. (2015). Vertebrates and Invertebrates of European Cities:Selected Non-Avian Fauna. Springer. p. 124. ISBN 978-1-4939-1698-6.
40. Abe, Y.; Bignell, David Edward; Higashi, T. (2014). Termites: Evolution, Sociality, Symbioses, Ecology. Springer. p. 437. ISBN 978-94-017-3223-9.
41. "Microhabitats". Australian National Botanic Gardens, Centre for Australian National Biodiversity Research (An Australian Government Initiative). Government of the Commonwealth of Australia. Retrieved 18 May 2016.
42. "Woodlands & Biodiversity". Offwell Woodland & Wildlife Trust. Retrieved 18 May 2016.
43. Lewis, E.E.; Campbell, J.F.; Sukhdeo, M.V.K. (2002). The Behavioural Ecology of Parasites. CABI. p. 183. ISBN 978-0-85199-754-4.
44. Pung, Oscar J.; Burger, Ashley R.; Walker, Michael F.; Barfield, Whitney L.; Lancaster, Micah H.; Jarrous, Christina E. (2009). "In vitro cultivation of Microphallus turgidus (Trematoda: Microphallidae) from metacercaria to ovigerous adult with continuation of the life cycle in the laboratory". Journal of Parasitology. 95 (4): 913–919. doi:10.1645/ge-1970.1. JSTOR 27735680.

45. Gorman, James (6 February 2013). "Bacteria Found Deep Under Antarctic Ice, Scientists Say". New York Times. Retrieved 18 May 2016.
46. Choi, Charles Q. (17 March 2013). "Microbes Thrive in Deepest Spot on Earth". LiveScience. Retrieved 18 May 2016.
47. Oskin, Becky (14 March 2013). "Intraterrestrials: Life Thrives in Ocean Floor". LiveScience. Retrieved 18 May 2016.
48. Schultz, Steven (13 December 1999). "Two miles underground". Princeton Weekly Bulletin. Archived from the original on 13 January 2016.
49. Chang, Kenneth (12 September 2016). "Visions of Life on Mars in Earth's Depths". New York Times. Retrieved 12 September 2016.
50. Cole, Bob (March 2008). "Appendix 6". The SAA BUhlmann DeeP-Stop System Handbook. Sub-Aqua Association. pp. vi–1. ISBN 978-0-9532904-8-2.
51. "The Deep Sea". MarineBio Conservation Society. 29 December 2011. Retrieved 19 May 2016.
52. "What does it take to live at the bottom of the ocean?". BBC Earth. 2016. Retrieved 19 May 2016.
53. "A hydrothermal vent forms when seawater meets hot magma". Ocean facts. National Ocean Service. 11 January 2013. Retrieved 20 May 2016.
54. "Hydrothermal Vent Creatures". Ocean Portal. Smithsonian National Museum of Natural History. Retrieved 20 May 2016.
55. Desbruyères, Daniel; Segonzac, Michel (1997). Handbook of Deep-sea Hydrothermal Vent Fauna. Editions Quae. p. 9. ISBN 978-2-905434-78-4.
56. Womack, Ann M.; Bohannan, Brendan J.M.; Green, Jessica L. (2010). "Biodiversity and biogeography of the atmosphere". Philiosophical Transactions of the Royal Society B. 365 (1558): 3645–3653. doi:10.1098/rstb.2010.0283.
57. Schulze-Makuch, Dirk; Haque, Shirin; Resendes de Sousa Antonio, Marina; Ali, Denzil; Hosein, Riad; Song, Young C.; Yang, Jinshu; Zaikova, Elena; Beckles, Denise M.; Guinan, Edward; Lehto, Harry J.; Hallam, Steven J. (2011). "Microbial Life in a Liquid Asphalt Desert". Astrobiology. 11 (3): 241–258. Bibcode:2011AsBio..11..241S. doi:10.1089/ast.2010.0488.
58. "Petroleum fly". Grzimek's Animal Life Encyclopedia. Volume 3: Insects (2nd ed.). The Gale Group. 2004. p. 367. ISBN 0-7876-5779-4.
59. McGregor, G.B.; Rasmussen, J.P. (2008). "Cyanobacterial composition of microbial mats from an Australian thermal spring: a polyphasic evaluation". FEMS Microbial Ecology. 63 (1): 23–35. doi:10.1111/j.1574-6941.2007.00405.x. PMID 18081588.
60. Hsing, Pen-Yuan (18 October 2010). "Gas-powered Circle of Life: Succession in a Deep-sea Ecosystem". Lophelia II 2010. NOAA. Retrieved 22 May 2016.
61. Gostincar, C.; Grube, M.; De Hoog, S.; Zalar, P.; Gunde-Cimerman, N. (2010). "Extremotolerance in fungi: Evolution on the edge". FEMS Microbiology Ecology. 71 (1): 2–11. doi:10.1111/j.1574-6941.2009.00794.x. PMID 19878320.

62. Takeuchi, Nozomu (2014). "Snow algae on Alaskan glaciers". Retrieved 22 May 2016.
63. Lindenmayer, David B.; Fischer, Joern (2013). Habitat Fragmentation and Landscape Change: An Ecological and Conservation Synthesis. Island Press. pp. 1–10. ISBN 978-1-59726-606-2.
64. Miller, G. Tyler; Spoolman, Scott (2008). Living in the Environment: Principles, Connections, and Solutions. Cengage Learning. pp. 193–195. ISBN 978-0-495-55671-8.
65. Holland, Matthew D.; Hastings, Alan (2008). "Strong effect of dispersal network structure on ecological dynamics". Nature. 456: 792–794. doi:10.1038/nature07395.
66. Brooks, M.L.; D'Antonio, C.M.; Richardson, D.M.; Grace, J.B.; Keeley, J.E.; DiTomaso, J.M.; Hobbs, R.J.; Pellant, M.; Pyke, D. (2004). "Effects of invasive alien plants on fire". BioScience. 54 (7): 677–688. doi:10.1641/0006-3568(2004)054[0677:EOIAPO]2.0.CO;2.
67. Lawrence, John M. (2013). Sea Urchins: Biology and Ecology. Academic Press. pp. 196–202. ISBN 978-0-12-397213-2.
68. de Klemm, Cyrille (1997). Comparative Analysis of the Effectiveness of Legislation for the Protection of Wild Flora in Europe. Council of Europe. pp. 65–70. ISBN 978-92-871-3429-5.
69. "Convention on the Conservation of Migratory Species of Wild Animals". UNEP/CMS Secretariat. Retrieved 7 July 2016.
70. Endangered Wildlife and Plants of the World. Marshall Cavendish. 2001. p. 750. ISBN 978-0-7614-7200-1.
71. Honey, Martha (2008). Ecotourism and Sustainable Development: Who Owns Paradise?. Island Press. p. 33. ISBN 1-59726-125-4.
72. Rio Guanajibo River Basin Flood Protection Project, Mayaguez, Hormigueros, San German: Environmental Impact Statement. 1994. p. 94.
73. "Mount Diablo Review" (pdf). Mount Diablo Interpretive Association. Autumn 2007. Retrieved 15 October 2008.
74. "1970 distribution of yellow starthistle in the U.S.". Yellow Starthistle Information. UCD.
75. "Invasive Mussels". National Wildlife Federation. Retrieved 29 June 2016.
76. Theel, Heather J.; Dibble, Eric D.; Madsen, John D. (2008). "Differential influence of a monotypic and diverse native aquatic plant bed on a macroinvertebrate assemblage; an experimental implication of exotic plant induced habitat". Hydrobiologia. 600: 77–87. doi:10.1007/s10750-007-9177-z.
77. "Definition of terms", Non-native Species Secretariat, accessed August 31, 2014.
78. "A neutral terminology to define 'invasive species'", Robert I. Colautti et al., Diversity and Distributions, (Diversity Distrib.) (2004) 10, 135–141.
79. Exotic Pest Plant Council. 'Exotic Pest Plants of Greatest Ecological Concern

in California' accessed 4/10/2010.

80. (September 21, 2006). National Invasive Species Information Center - What is an Invasive Species?. United States Department of Agriculture: National Agriculture Library. Retrieved on September 1, 2007.
81. USA (1999). "Executive Order 13112 of February 3, 1999: Invasive Species". Federal Register. 64 (25): 6183–6186.
82. "Communication From The Commission To The Council, The European Parliament, The European Economic And Social Committee And The Committee Of The Regions Towards An EU Strategy On InvasFpollive Species" (PDF). Retrieved 2011-05-17.
83. Tsiamis, Konstantinos; Gervasini, Eugenio; D'Amico, Fabio; Deriu, Ivan; Katsanevakis, Stelios; Crocetta, Fabio; Zenetos, Argyro; Arianoutsou, Margarita; Backeljau, Thierry; Bariche, Michel; Bazos, Ioannis; Bertaccini, Assunta; Brundu, Giuseppe; Carrete, Martina; Çinar, Melih; Curto, Giovanna; Faasse, Marco; Justine, Jean-Lou; Király, Gergely; Langer, Martin; Levitt, Ya'arit; Panov, Vadim; Piraino, Stefano; Rabitsch, Wolfgang; Roques, Alain; Scalera, Riccardo; Shenkar, Noa; Sîrbu, Ioan; Tricarico, Elena; Vannini, Andrea; Vøllestad, Leif Asbjørn; Zikos, Andreas; Cardoso, Ana Cristina (2016). "The EASIN Editorial Board: quality assurance, exchange and sharing of alien species information in Europe" (PDF). Management of Biological Invasions. 7 (4): 321–328. doi:10.3391/mbi.2016.7.4.02. ISSN 1989-8649.
84. Exotic Pest Plant Council. p. 1. accessed 4/10/2010.
85. Kolar, C.S.; D.M. Lodge (2001). "Progress in invasion biology: predicting invaders". Trends in Ecology & Evolution. 16 (4): 199–204. doi:10.1016/S0169-5347(01)02101-2. PMID 11245943.
86. Thebaud, C.; A.C. Finzi, L. Affre, M. Debussche, J. Escarre (1996). "Assessing why two introduced Conyza differ in their ability to invade Mediterranean old fields". Ecology. 77 (3): 791–804. doi:10.2307/2265502. JSTOR 2265502.
87. Reichard, S.H.; C. W. Hamilton (1997). "Predicting invasions of woody plants introduced into North America". Conservation Biology. 11 (1): 193–203. doi:10.1046/j.1523-1739.1997.95473.x.
88. Williams, J.D.; G. K. Meffe (1998). "Nonindigenous Species". Status and Trends of the Nation's Biological Resources. Reston, Virginia: United States Department of the Interior, Geological Survey. 1.
89. Ewell, J.J.; D.J. O'Dowd, J. Bergelson, C.C. Daehler, C.M. D'Antonio, L.D. Gomez, D.R. Gordon, R.J. Hobbs, A. Holt, K.R. Hopper, C.E. Hughes, M. LaHart, R.R.B. Leakey, W.G. Wong, L.L. Loope, D.H. Lorence, S.M. Louda, A.E. Lugo, P.B. McEvoy, D.M. Richardson, and P.M. Vitousek (1999). "Deliberate introductions of species: Research needs - Benefits can be reaped, but risks are high". BioScience. 49 (8): 619–630. doi:10.2307/1313438. JSTOR 1313438.
90. Tilman, D. (2004). "Niche tradeoffs, neutrality, and community structure: A stochastic theory of resource competition, invasion, and community assembly". Proceedings of the National Academy of Sciences. 101 (30): 10854–10861.

doi:10.1073/pnas.0403458101. PMC 503710 . PMID 15243158.

91. Verling, E.; G.M. Ruiz, L.D. Smith, B. Galil, A.W. Miller, and K.R. Murphy (2005). “Supply-side invasion ecology: characterizing propagule pressure in coastal ecosystems”. Proceedings of the Royal Society B. 272 (1569): 1249–1256. doi:10.1098/rspb.2005.3090. PMC 1564104 . PMID 16024389.

92. Stohlgren, T.J.; D. Binkley, G.W. Chong, M.A. Kalkhan, L.D. Schell, K.A. Bull, Y. Otsuki, G. Newman, M. Bashkin, and Y. Son; Chong, Geneva W.; Kalkhan, Mohammed A.; Schell, Lisa D.; Bull, Kelly A.; Otsuki, Yuka; Newman, Gregory; Bashkin, Michael; Son, Yowhan (1999). “Exotic plant species invade hot spots of native plant diversity”. Ecological Monographs. 69: 25–46. doi:10.1890/0012-9615(1999)069[0025:EPSIHS]2.0.CO;2.

93. Sax, D.F.; S. D. Gaines and J. H. Brown (2002). “Species Invasions Exceed Extinctions on Islands Worldwide: A Comparative Study of Plants and Birds”. American Naturalist. 160 (6): 766–783. doi:10.1086/343877. PMID 18707464.

94. Huenneke, L.; S. Hamburg; R. Koide; H. Mooney; P. Vitousek (1990). “Effects of soil resources on plant invasion and community structure in California (USA) serpentine grassland”. Ecology. 71 (2): 478–491. doi:10.2307/1940302. JSTOR 1940302.

95. Herrera, Ileana; Ferrer-Paris, José R.; Hernández-Rosas, José I.; Nassar, Jafet M. (2016). “Impact of two invasive succulents on native-seedling recruitment in Neotropical arid environments” (PDF). Journal of Arid Environments. 132: 15–25. doi:10.1016/j.jaridenv.2016.04.007. Retrieved 6 May 2016.

96. Hierro, J.L.; R.M. Callaway (2003). “Allelopathy and exotic plant invasion”. Plant and Soil. 256 (1): 29–39. doi:10.1023/A:1026208327014.

97. Vivanco, J.M.; Bais, H.P.; Stermitz, F.R.; Thelen, G.C.; Callaway, R.M. (2004). “Biogeographical variation in community response to root allelochemistry: Novel weapons and exotic invasion”. Ecology Letters. 7 (4): 285–292. doi:10.1111/j.1461-0248.2004.00576.x.

98. Brooks, M.L.; D’Antonio, C.M.; Richardson, D.M.; Grace, J.B.; Keeley, J.E.; DiTomaso, J.M.; Hobbs, R.J.; Pellant, M.; Pyke, D. (2004). “Effects of invasive alien plants on fire”. BioScience. 54 (7): 677–688. doi:10.1641/0006-3568(2004)054[0677:EOIAPO]2.0.CO;2.

99. Silver Botts, P.; Patterson, B.A.; Schlosser, D. (1996). “Zebra mussel effects on benthic invertebrates: Physical or biotic?”. Journal of the North American Benthological Society (15): 179–184.

100. Byers, J.E. (2002). “Impact of non-indigenous species on natives enhanced by anthropogenic alteration of selection regimes”. Oikos. 97 (3): 449–458. doi:10.1034/j.1600-0706.2002.970316.x.

101. Davis, M.A.; Grime, J.P.; Thompson, K. (2000). “Fluctuating resources in plant communities: A general theory of invisibility”. Journal of Ecology. 88 (3): 528–534. doi:10.1046/j.1365-2745.2000.00473.x.

102. Xu, Cheng-Yuan; Tang, Shaoqing; Fatemi, Mohammad; Gross, Caroline L.; Julien, Mic H.; Curtis, Caitlin; van Klinken, Rieks D. (2015-09-01). “Population

structure and genetic diversity of invasive Phyla canescens: implications for the evolutionary potential". Ecosphere. 6 (9): 1–21. doi:10.1890/ES14-00374.1. ISSN 2150-8925.

103. Prentis, Peter. "Adaptive evolution in invasive species". Trends in Plant Science. 13 (6): 288–294. doi:10.1016/j.tplants.2008.03.004.

104. Lee, Carol Eunmi. "Evolutionary genetics of invasive species". Trends in Ecology & Evolution. 17 (8): 386–391. doi:10.1016/s0169-5347(02)02554-5.

105. Zenni, R.D. "Adaptive Evolution and Phenotypic Plasticity During Naturalization and Spread of Invasive Species: Implications for Tree Invasion Biology". Biological Invasions. 16 (3): 635–644. doi:10.1007/s10530-013-0607-8.

106. Elton, C.S. (2000) [1958]. The Ecology of Invasions by Animals and Plants. Foreword by Daniel Simberloff. Chicago: University of Chicago Press. p. 196. ISBN 0-226-20638-6.

107. Stohlgren, T.J.; Binkley, D.; Chong, G.W.; Kalkhan, M.A.; Schell, L.D.; Bull, K.A.; Otsuki, Y.; Newman, G.; Bashkin, M.; Son, Y. (1999). "Exotic plant species invade hot spots of native plant diversity". Ecological Monographs. 69: 25–46. doi:10.1890/0012-9615(1999)069[0025:EPSIHS]2.0.CO;2.

108. Byers, J.E.; E.G. Noonburg (2003). "Scale dependent effects of biotic resistance to biological invasion". Ecology. 84 (6): 1428–1433. doi:10.1890/02-3131.

109. Levine, J. M. (2000). "Species diversity and biological invasions: Relating local process to community pattern". Science. 288 (5467): 852–854. doi:10.1126/science.288.5467.852. PMID 10797006.

110. Williams, J.D.; G. K. Meffe (1998). "Nonindigenous Species". Status and Trends of the Nation's Biological Resources. Reston, Virginia: United States Department of the Interior, Geological Survey. 1.

111. Stachowicz, J.J.; D. Tilman (2005). "Species invasions and the relationships between species diversity, community saturation, and ecosystem functioning". In D.F. Sax; J.J. Stachowicz; S.D. Gaines. Species Invasions: Insights into Ecology, Evolution, and Biogeography. Sunderland, Massachusetts: Sinauer Associates. ISBN 0-87893-811-7.

112. Cassey, P; T.M. Blackburn; R.P. Duncan; S.L. Chown (2005). "Concerning Invasive Species: Reply to Brown and Sax". Austral Ecology. 30 (4): 475–480. doi:10.1111/j.1442-9993.2005.01505.x.

113. Matisoo-Smith, E.; R.M. Roberts; G.J. Irwin; J.S. Allen; D. Penny; D.M. Lambert (1998). "Patterns of prehistoric human mobility in Polynesia indicated by mtDNA from the Pacific rat". Proceedings of the National Academy of Sciences of the United States of America. 95 (25): 15145–15150. doi:10.1073/pnas.95.25.15145. PMC 24590 . PMID 9844030.

114. Molnar, Jennifer L; Gamboa, Rebecca L; Revenga, Carmen; Spalding, Mark D (2008). "Assessing the global threat of invasive species to marine biodiversity". Frontiers in Ecology and the Environment. 6 (9): 485–492. doi:10.1890/070064. ISSN 1540-9295.

115. Xu, Jian; Wickramarathne, Thanuka L.; Chawla, Nitesh V.; Grey, Erin K.; Steinhaeuser, Karsten; Keller, Reuben P.; Drake, John M.; Lodge, David M. (2014). "Improving management of aquatic invasions by integrating shipping network, ecological, and environmental data". Proceedings of the 20th ACM SIGKDD international conference on Knowledge discovery and data mining - KDD ‹14: 1699–1708. doi:10.1145/2623330.2623364.

116. F. Moretzsohn; J.A. Sánchez Chávez; J.W. Tunnell, Jr. (eds.). "Invasive Species". GulfBase: Resource Database for Gulf of Mexico Research. Harte Research Institute for Gulf of Mexico Studies at Texas A&M University-Corpus Christi. Retrieved March 19, 2013.

117. Aquatic invasive species. A Guide to Least-Wanted Aquatic Organisms of the Pacific Northwest. 2001. University of Washington

118. jsonline.com: "Noxious cargo" 31 Oct 2005

119. Leung, B.; N.E. Mandrak (2007). "The risk of establishment of aquatic invasive species: joining invasibility and propagule pressure". Proceedings of the Royal Society B. 274 (1625): 2733–2739. doi:10.1098/rspb.2007.0841. PMC 2275890 . PMID 17711834.

120. "Our Invaluable Invertebrate Collections". Ars.usda.gov. Retrieved 2011-05-17.

121. Grosholz, E.D. (2005). "Recent biological invasion may hasten invasional meltdown by accelerating historical introductions". Proceedings of the National Academy of Sciences. 102 (4): 1088–1091. doi:10.1073/pnas.0308547102. PMC 545825 . PMID 15657121.

122. Mack, R.; D. Simberloff; W.M. Lonsdale; H. Evans; M. Clout; F.A. Bazzazf (2000). "Biotic invasions: Causes, epidemiology, global consequences, and control". Ecological Applications. 10 (3): 689–710. doi:10.1890/1051-0761(2000)010[0689:BICEGC]2.0.CO;2.

123. Hawkes, C.V.; I.F. Wren; D.J. Herman; M.K. Firestone (2005). "Plant invasion alters nitrogen cycling by modifying the soil nitrifying community". Ecology Letters. 8 (9): 976–985. doi:10.1111/j.1461-0248.2005.00802.x.

124. Rhymer, J. M.; Simberloff, D. (1996). "Extinction by hybridization and introgression". Annual Review of Ecology and Systematics. 27 (1): 83–109. doi:10.1146/annurev.ecolsys.27.1.83.

125. Ayres, D.; et al. (2004). "Spread of exotic cordgrasses and hybrids (Spartina sp.) in the tidal marshes of San Francisco Bay, California". USA Biological Invasions. 6 (2): 221–231. doi:10.1023/B:BINV.0000022140.07404.b7.

126. Primtel, David (2005). "Update on the environmental and economic costs associated with alien-invasive species in the United States". Ecological Economics. 52: 273–288. doi:10.1016/j.ecolecon.2004.10.002.

127. Fei, S.; J. Philips; M. Shouse (2015). "Biogeomorphic Impacts of Invasive Species". The Annual Review of Ecology, Evolution, and Systematics. 45: 69–87. doi:10.1146/annurev-ecolsys-120213-091928.

128. Energy from vascular plant wastewater treatment systems - Eichhornia crassipes,

Spirodela lemna, Hydrocotyle ranunculoides, Pueraria lobata, biomass harvested for fuel production. By B.C. Wolverton and R.C. McDonald. 1981. Econ. Bot. 35(2):224-232. Cited in Handbook of Energy Crops. By J. Duke.

129. http://onlinelibrary.wiley.com/doi/10.1002/bbb.1539/abstract
130. Tom Pelton, Baltimore Sun, May 26, 2006.
131. Zayed, Amro; Constantin, Şerban A.; Packer, Laurence (2007-09-12). "Successful Biological Invasion despite a Severe Genetic Load". PLoS ONE. 2 (9): e868. doi:10.1371/journal.pone.0000868. ISSN 1932-6203. PMC 1964518 . PMID 17848999.
132. Adamson, Nancy Lee. An Assessment of Non-Apis Bees as Fruit and Vegetable Crop Pollinators in Southwest Virginia. Diss. 2011. Web. 15 Oct. 2015.
133. "http://www.scientificamerican.com/article/invasive-species-menu-of-a-world-class-chef/"
134. "http://www.theguardian.com/environment/world-on-a-plate/2015/feb/06/cooking-cant-solve-the-invasive-threat"
135. "http://news.yahoo.com/invasive-lionfish-kings-caribbean-may-met-match-011600208.html"
136. Pimentel, D.; R., Zuniga; Morrison, D (2005). "Update on the environmental and economic costs associated with alien-invasive species in the United States". Ecological Economics. 52 (3): 273–288. doi:10.1016/j.ecolecon.2004.10.002.
137. Simberloff, D. (2001). "Biological invasions - How are they affecting us, and what can we do about them?". Western North American Naturalist. 61: 308–315.
138. 2013. "Citrus Greening." Invasive Species Program - Pest Alerts. Clemson University - DPI. http://www.clemson.edu/public/regulatory/plant_industry/invasive_exotic_programs/Pest%20Alerts/citrus_greening.html. Accessed 24 May 2013.
139. Azevedo-Santos, V.M.; O. Rigolin-Sá; F.M. Pelicice (2011). "Growing, losing or introducing? Cage aquaculture as a vector for the introduction of non-native fish in Furnas Reservoir, Minas Gerais, Brazil". Neotropical Ichthyology. 9 (4): 915–9. doi:10.1590/S1679-62252011000400024. ISSN 1679-6225.
140. Liebhold, S.; et al. (2013). "A highly aggregated geographical distribution of forest pest invasions in the USA". Diversity and Distribution. 19 (9): 1208–1216. doi:10.1111/ddi.12112.
141. Oswalt, C.; et al. (2015). "A subcontinental view of forest plant invasions". NeoBiota. 24: 49–54. doi:10.3897/neobiota.24.8378.
142. Balsam woolly aphid Adelges piceae (Ratzeburg) ForestPests.org (March 3, 2005) Retrieved on September 1, 2007.
143. Scott E. Schlarbaum; Frederick Hebard; Pauline C. Spaine; Joseph C. Kamalay (1997). "Three American Tragedies: Chestnut Blight, Butternut Canker and Dutch Elm Disease". (originally published via: Proceedings: Exotic Pests of Eastern Forests; (1997 April 8–10); Nashville, TN. Tennessee Exotic Pest Plant Council: 45-54.). Southern Research Station, Forest Service, United States Department

of Agriculture. Retrieved June 22, 2012.

144. Alternative link and additional publication citation information: Tree Search, US Forest Service, USDA. http://www.treesearch.fs.fed.us/pubs/745

145. Rodger, Vikki; Stinson, Kristin; Finzi, Adrian (2008). "Ready or Not, Garlic Mustard Is Moving In: Alliaria petiolata as a Member of Eastern North American Forests". BioScience. 58: 5. doi:10.1641/b580510.

146. Eiswerth, M.E.; Darden, Tim D.; Johnson, Wayne S.; Agapoff, Jeanmarie; Harris, Thomas R. (2005). "Input-output modeling, outdoor recreation, and the economic impacts of weeds". Weed Science. 53: 130–137. doi:10.1614/WS-04-022R.

147. Eurasian Watermilfoil in the Great Lakes Region. GreatLakes.net. Retrieved on September 1, 2007.

148. Sin, Hans; Adam Radford (2007). "Coqui frog research and management efforts in Hawaii" (PDF). Managing Vertebrate Invasive Species: Proceedings of an International Symposium (G. W. Witmer, W. C. Pitt, K. A. Fagerstone, Eds). USDA/APHIS/WS, National Wildlife Research Center, Fort Collins, CO. Retrieved 2013-06-26.

149. Aquatic invasive species. A Guide to Least-Wanted Aquatic Organisms of the Pacific Northwest. 2001. University of Washington

150. Lanciotti, R. S.; Roehrig, J. T.; Deubel, V.; Smith, J.; Parker, M.; Steele, K.; Crise, B.; Volpe, K. E.; et al. (1999). "Origin of the West Nile virus responsible for an outbreak of encephalitis in the northeastern United States". Science. 286 (5448): 2333–2337. doi:10.1126/science.286.5448.2333. PMID 10600742.

151. Hallegraeff, G.M. (1998). "Transport of toxic dinoflagellates via ships' ballast water: Bioeconomic risk assessment and efficacy of possible ballast water management strategies". Marine Ecology Progress Series. 168: 297–309. doi:10.3354/meps168297.

152. Millennium Ecosystem Assessment (2005). "Ecosystems and Human Well-being: Biodiversity Synthesis" (PDF). World Resources Institute.

153. Odendaal, L. J.; Haupt, T. M.; Griffiths, C. L. (2008). "The alien invasive land snail Theba pisana in the West Coast National Park: Is there cause for concern?".". Koedoe. 50 (1): 93–98. doi:10.4102/koedoe.v50i1.153.

154. Mooney, HA; Cleland, EE (2001). "The evolutionary impact of invasive species". Proceedings of the National Academy of Sciences of the United States of America. 98 (10): 5446–51. doi:10.1073/pnas.091093398. PMC 33232 . PMID 11344292.

155. "Glossary: definitions from the following publication: Aubry, C., R. Shoal and V. Erickson. 2005. Grass cultivars: their origins, development, and use on national forests and grasslands in the Pacific Northwest. USDA Forest Service. 44 pages, plus appendices.; Native Seed Network (NSN), Institute for Applied Ecology, 563 SW Jefferson Ave, Corvallis, OR 97333, USA". Nativeseednetwork.org. Retrieved 2011-05-17.

156. Rhymer, Judith M.; Simberloff, Daniel. "Extinction by Hybridization and Introgression". Annual Review of Ecology and Systematics. 27: 83–109. doi:10.1146/annurev.ecolsys.27.1.83.

157. Genetic Pollution from Farm Forestry using eucalypt species and hybrids; A report for the RIRDC/L&WA/FWPRDC; Joint Venture Agroforestry Program; by Brad M. Potts, Robert C. Barbour, Andrew B. Hingston; September 2001; RIRDC Publication No 01/114; RIRDC Project No CPF - 3A; ISBN 0-642-58336-6 ISSN 1440-6845; Australian Government, Rural Industrial Research and Development Corporation]

158. Lockwood, Julie L.; Martha F. Hoopes; Michael P. Marchetti (2007). Invasion Ecology (PDF). Blackwell Publishing. p. 7. Retrieved 21 January 2014.

159. Lowry, E; Rollinson, EJ; Laybourn, AJ; Scott, TE; Aiello-Lammens, ME; Gray, SM; Mickley, J; Gurevitch, J (2012). "Biological invasions: A field synopsis, systematic review, and database of the literature". Ecology and evolution. 3 (1): 182–96. doi:10.1002/ece3.431. PMC 3568853 . PMID 23404636.

160. Colautti, Robert I.; Hugh J. MacIsaac (2004). "A neutral terminology to define ‹invasive' species" (PDF). Diversity and Distributions. 10 (2): 135–141. doi:10.1111/j.1366-9516.2004.00061.x. Retrieved 2007-09-01.

CHAPTER 6

Conservation: Ethical Use and Protection of Valuable Resources

Chapter Contents

Usage, improvement, and protection of human and natural resources in a wise manner, ensuring derivation of their highest economic and social benefits on a continuing or long-term basis. Conservation is achieved through alternative technologies, recycling, and reduction in waste and spoilage, and (unlike preservation) implies consumption of the conserved resources.

INTRODUCTION

Conservation is the ethical use and protection of valuable resources, such as trees, minerals, wildlife, water, and others. It focuses on maintaining the natural world in order to protect the sources of resources. Conservation may refer to:

- Conservation (ethic) of biodiversity, environment, and natural resources, including protection and management
- Conservation-restoration of cultural heritage, protection and restoration of cultural heritage, including works of art and architecture, as well as archaeological and historical artifacts

- Conservation law, measurable property of isolated physical system that does not change as the system evolves, including conservation of energy, mass, momentum, electric charge, subatomic particles, and fundamental symmetries

Subtopics

Conservation of Biodiversity, Environment, and Natural Resources

- Conservation biology, the science of protection and management of biodiversity
- Conservation (ethic), an ethic of use, protection, and management of the environment and natural resources
- Conservation finance, the practice of raising and managing capital to support land, water, and resource conservation
- Conservation genetics, an interdisciplinary science that applies genetic methods to conservation and restoration of biodiversity
- Conservation movement, political, environmental, or social movement that seeks to protect natural resources, including biodiversity and habitat
- Conservation organization an organization dedicated to protection and management of the environment or natural resources
- Conservationist (Conservation movement), a person who advocates for conservation of the environment and natural resources
- Conservation of natural resources
- Energy conservation, management of energy resources
- Habitat conservation, land management practices to protect, manage, and restore habitat
- Marine conservation, protection and management of ocean and sea ecosystems
- Soil conservation, management of soils to prevent erosion, depletion, and contamination
- Water conservation, management of fresh water as a sustainable resource
- Wetland conservation, management to protect and restore wetlands
- Wildlife conservation, protection of endangered species and their habitat
- Wildlife management, management to conserve wild species and their habitats

Conservation of Cultural Heritage

Conservation-restoration of cultural heritage

- Architectural conservation
- Book and paper conservation
- Conservation science (cultural heritage)

- Frame conservation
- Object conservation (archaeological, ethnographic, and sculptural)
- Photograph conservation
- Preventive conservation
- Stained glass conservation

Physical Laws of Conservation

- Conservation of energy
- Conservation of mass
- Conservation of mass-energy
- Conservation of linear momentum
- Conservation of angular momentum
- Charge conservation
- Conservation of color charge
- Conservation of weak isospin
- Conservation of probability (Probability current)
- CPT symmetry (combining charge, parity, and time conjugation)
- Lorentz symmetry

CONSERVATION GENETICS

Conservation genetics is an interdisciplinary subfield of Population Genetics that aims to understand the dynamics of genes in populations principally to avoid extinction. Therefore, it applies genetic methods to the conservation and restoration of biodiversity. Researchers involved in conservation genetics come from a variety of fields including population genetics, molecular ecology, biology, evolutionary biology, and systematics. Genetic diversity is one of the three fundamental levels of biodiversity, so it is directly important in conservation. Genetic variability influences both the health and long-term survival of populations because decreased genetic diversity has been associated with reduced fitness, such as high juvenile mortality, diminished population growth, reduced immunity, and ultimate to higher extinction risk.

Genetic Diversity

Genetic diversity is the variability of genes in a species. A number of means can express the level of genetic diversity: observed heterozygosity, expected heterozygosity, the mean number of alleles per locus, or the percentage of polymorphic loci.

Importance of Genetic Diversity

Since the genes encode the information of the phenotype, genetic diversity determines the potential fitness of a population and ultimately its long-term persistence. Extinction risk has been associated with low genetic diversity because several researchers has documented reduced fitness in population with low genetic diversity. For example, low genetic diversity as low heterozigosity has been associated with low juvenile survival, reduced population growth, low body size, diminished adult lifespan.

The importance of genetic diversity has many reasons. For example, one can consider the probability of within any given population that at least one individual or a potential reproductive pair have the genetic composition encoding a phenotype capable of survival an environment event. If 1 the individuals are nearly identical. If new pressures (such as environmental disasters) occur, a population with high genetic diversity has a greater chance of having at least some individuals with a genetic makeup that allows them to survive. If genetic diversity is very low, none of the individuals in a population may have the characteristics needed to cope with the new environmental conditions. Such a population could be suddenly wiped out.

The genetic diversity of a species is always open to change. No matter how many variants of a gene are present in a population today, only the variants that survive in the next generation can contribute to species diversity in the future. Once gene variants are lost, they cannot be recovered.

Contributors to Extinction

- Inbreeding and inbreeding depression.
- The accumulation of deleterious mutations
- A decrease in frequency of heterozygotes in a population, or heterozygosity, which decreases a species' ability to evolve to deal with change in the environment.
- Outbreeding depression
- Fragmented populations
- Taxonomic uncertainties, which can lead to a reprioritization of conservation efforts
- Genetic drift as the main evolutionary process, instead of natural selection
- Management units within species

Techniques

Specific genetic techniques are used to assess the genomes of a species regarding specific conservation issues as well as general population structure. This analysis can be done in two ways, with current DNA of individuals or historic DNA.

Techniques for analysing the differences between individuals and populations include:

- Alloenzymes
- Random fragment length polymorphisms
- Amplified fragment length polymorphisms
- Random amplification of polymorphic DNA
- Single strand conformation polymorphism
- Minisatellites
- Microsatellites
- Single-nucleotide polymorphisms
- DNA Sequence

These different techniques focus on different variable areas of the genomes within animals and plants. The specific information that is required determines which techniques are used and which parts of the genome are analysed. For example, mitochondrial DNA in animals has a high substitution rate, which makes it useful for identifying differences between individuals. However, it is only inherited in the female line, and the mitochondrial genome is relatively small. In plants, the mitochondrial DNA has very high rates of structural mutations, so is rarely used for genetic markers, as the chloroplast genome can be used instead. Other sites in the genome that are subject to high mutation rates such as the major histocompatibility complex, and the microsatellites and minisatellites are also frequently used.

These techniques can provide information on long-term conservation of genetic diversity and expound demographic and ecological matters such as taxonomy.

Another technique is using historic DNA for genetic analysis. Historic DNA is important because it allows geneticists to understand how species reacted to changes to conditions in the past. This is a key to understanding the reactions of similar species in the future.

Techniques using historic DNA include looking at preserved remains found in museums and caves. Museums are used because there is a wide range of species that are available to scientists all over the world. The problem with museums is that, historical perspectives are important because understanding how species reacted to changes in conditions in the past is a key to understanding reactions of similar species in the future. Evidence found in caves provides a longer perspective and does not disturb the animals.

Another technique that relies on specific genetics of an individual is noninvasive monitoring, which uses extracted DNA from organic material that an individual leaves behind, such as a feather. This too avoids disrupting the animals and can provide information about the sex, movement, kinship and diet of an individual.

Other more general techniques can be used to correct genetic factors that lead to extinction and risk of extinction. For example, when minimizing inbreeding and increasing genetic variation multiple steps can be taken. Increasing heterozygosity through immigration, increasing the generational interval through cryopreservation or breeding from older animals, and increasing the effective population size through equalization of family size all helps minimize inbreeding and its effects. Deleterious alleles arise through mutation, however certain recessive ones can become more prevalent due to inbreeding. Deleterious mutations that arise from inbreeding can be removed by purging, or natural selection. Populations raised in captivity with the intent of being reintroduced in the wild suffer from adaptations to captivity.

Inbreeding depression, loss of genetic diversity, and genetic adaptation to captivity are disadvantageous in the wild, and many of these issues can be dealt with through the aforementioned techniques aimed at increasing heterozygosity. In addition creating a captive environment that closely resembles the wild and fragmenting the populations so there is less response to selection also help reduce adaptation to captivity.

Solutions to minimize the factors that lead to extinction and risk of extinction often overlap because the factors themselves overlap. For example, deleterious mutations are added to populations through mutation, however the deleterious mutations conservation biologists are concerned with are ones that are brought about by inbreeding, because those are the ones that can be taken care of by reducing inbreeding. Here the techniques to reduce inbreeding also help decrease the accumulation of deleterious mutations.

Applications

These techniques have wide ranging applications. One application of these specific molecular techniques is in defining species and sub-species of salmonids. Hybridization is an especially important issue in salmonids and this has wide ranging conservation, political, social and economic implications. In Cutthroat Trout mtDNA and alloenzyme analysis, hybridization between native and non-native species was shown to be one of the major factors contributing to the decline in their populations. This led to efforts to remove some hybridized populations so native populations could breed more readily. Cases like these impact everything from the economy of local fishermen to larger companies, such as timber. Specific molecular techniques led to a closer analysis of taxonomic relationships, which is one factor that can lead to extinctions if unclear.

Implications

New technology in conservation genetics has many implications for the future of conservation biology. At the molecular level, new technologies are advancing. Some of these techniques include the analysis of minisatellites and MHC. These molecular techniques have wider effects from clarifying taxonomic relationships, as in the previous example, to determining the best individuals to reintroduce to a population for recovery by determining kinship. These effects then have consequences that reach even further. Conservation of species has implications for humans in the economic, social, and political realms. In the biological realm increased genotypic diversity has been shown to help ecosystem recovery, as seen in a community of grasses which was able to resist disturbance to grazing geese through greater genotypic diversity. Because species diversity increases ecosystem function, increasing biodiversity through new conservation genetic techniques has wider reaching effects than before.

A short list of studies a conservation geneticist may research include:

- Phylogenetic classification of species, subspecies, geographic races, and populations, and measures of phylogenetic diversity and uniqueness.
- Identifying hybrid species, hybridization in natural populations, and assessing the history and extent of introgression between species.
- Population genetic structure of natural and managed populations, including identification of Evolutionary Significant Units (ESUs) and management units for conservation.
- Assessing genetic variation within a species or population, including small or endangered populations, and estimates such as effective population size (Ne).
- Measuring the impact of inbreeding and outbreeding depression, and the relationship between heterozygosity and measures of fitness (see Fisher›s fundamental theorem of natural selection).
- Evidence of disrupted mate choice and reproductive strategy in disturbed populations.
- Forensic applications, especially for the control of trade in endangered species.
- Practical methods for monitoring and maximizing genetic diversity during captive breeding programs and re-introduction schemes, including mathematical models and case studies.
- Conservation issues related to the introduction of genetically modified organisms.
- The interaction between environmental contaminants and the biology and health of an organism, including changes in mutation rates and adaptation to local changes in the environment (e.g. industrial melanism).
- New techniques for noninvasive genotyping.

REFERENCES

1. Avise, John C & Hamrick James L (eds.). Conservation Genetics. Springer. ISBN 0-412-05581-3.
2. Frankham, Richard (2003). "Genetics and Conservation Biology". Comptes Rendus Biologies. 326: S22–S29. doi:10.1016/S1631-0691(03)00023-4. PMID 14558445.
3. Allendorf, F.W. and G. Luikart (2007). Conservation and the Genetics of Populations. Wiley-Blackwell. p. 642.

CHAPTER 7

Conservation Biology

Chapter Contents

Conservation biology incorporates many fields of scientific research and environmental management studies. Biologists study evolution, ecology, and Earth science to better understand how populations of organisms react to changes in their environment. Conservation biology is important in identifying and countering many current threats to ecosystems and biodiversity. Scientists often focus on educating the public about the dangers of global warming, deforestation, and pollution, as well as advocating the benefits of sustainable practices.

INTRODUCTION

Conservation biology is the management of nature and of Earth's biodiversity with the aim of protecting species, their habitats, and ecosystems from excessive rates of extinction and the erosion of biotic interactions. It is an interdisciplinary subject drawing on natural and social sciences, and the practice of natural resource management.

The conservation ethic is based on the findings of conservation biology.

Efforts are made to preserve the natural characteristics of Hopetoun Falls, Australia, without access by visitors being affected.

Origination of Conservation Biology

The term conservation biology and its conception as a new field originated with the convening of "The First International Conference on Research in Conservation Biology" held at the University of California, San Diego in La Jolla, California in 1978. led by American biologists Bruce A. Wilcox and Michael E. Soulé with a group of leading university and zoo researchers and conservationists including Kurt Benirschke, Sir Otto Frankel, Thomas E. Lovejoy, and Jared M. Diamond. The meeting was prompted by the concern over tropical deforestation, disappearing species, eroding genetic diversity within species. The conference and proceedings that resulted sought to initiate the bridging of a gap between theory in ecology and evolutionary genetics on the one hand and conservation policy and practice on the other. Conservation biology and the concept of biological diversity (biodiversity) emerged together, helping crystallize the modern era of conservation science and policy. The inherent multidisciplinary basis for conservation biology has led to new subdisciplines including conservation social science, conservation behavior, and conservation physiology. It stimulated further development of conservation genetics which Otto Frankel had originated first but is now often considered a subdiscipline as well.

Description

The rapid decline of established biological systems around the world means that conservation biology is often referred to as a "Discipline with a deadline". Conservation biology is tied closely to ecology in researching the population ecology (dispersal, migration, demographics, effective population size, inbreeding depression, and minimum population viability) of rare or endangered species. Conservation biology is concerned with phenomena that affect the maintenance, loss, and restoration of biodiversity and the science of sustaining evolutionary processes that engender genetic, population, species, and ecosystem diversity. The concern stems from estimates suggesting that up to 50% of all species on the planet will disappear within the next 50 years, which has contributed to poverty, starvation, and will reset the course of evolution on this planet.

Conservation biologists research and educate on the trends and process of biodiversity loss, species extinctions, and the negative effect these are having on our capabilities to sustain the well-being of human society. Conservation biologists work in the field and office, in government, universities, non-profit organizations and industry. The topics of their research are diverse, because this is an interdisciplinary network with professional alliances in the biological, as well as social sciences. Those dedicated to the cause and profession advocate for a global response to the current biodiversity crisis based on morals, ethics, and scientific reason. Organizations and citizens are responding to the biodiversity crisis through conservation action plans that direct research, monitoring, and educational programs that engage concerns at local through global scales.

Natural Resource Conservation

Conscious efforts to conserve and protect *global* biodiversity are recent phenomena. Natural resource conservation, however, has a history that extends prior to the age of conservation. Resource ethics grew out of necessity through direct relations with nature. Regulation or communal restraint became necessary to prevent selfish motives from taking more than could be locally sustained, therefore compromising the long-term supply for the rest of the community. This social dilemma with respect to natural resource management is often called the "Tragedy of the Commons".

From this principle, conservation biologists can trace communal resource based ethics throughout cultures as a solution to communal resource conflict. For example, the Alaskan Tlingit peoples and the Haida of the Pacific Northwest had resource boundaries, rules, and restrictions among clans with respect to the fishing of sockeye salmon. These rules were guided by clan elders who knew lifelong details of each river and stream they managed. There are numerous examples in history where cultures have followed rules, rituals, and organized practice with respect to communal natural resource management.

Conservation ethics are also found in early religious and philosophical writings. There are examples in the Tao, Shinto, Hindu, Islamic and Buddhist traditions. In Greek philosophy, Plato lamented about pasture land degradation: "What is left now is, so to say, the skeleton of a body wasted by disease; the rich, soft soil has been carried off and only the bare framework of the district left." In the bible, through Moses, God commanded to let the land rest from cultivation every seventh year. Before the 18th century, however, much of European culture considered it a pagan view to admire nature. Wilderness was denigrated while agricultural development was praised. However, as early as AD 680 a wildlife sanctuary was founded on the Farne Islands by St. Cuthbert in response to his religious beliefs.

Early Naturalists

White gyrfalcons drawn by John James Audubon

Natural history was a major preoccupation in the 18th century, with grand expeditions and the opening of popular public displays in Europe and North America. By 1900, there were 150 natural history museums in Germany, 250 in Great Britain, 250 in the United States, and 300 in France. Preservationist or conservationist sentiments are a development of the late 18th to early 20th centuries.

Before Charles Darwin set sail on the HMS Beagle, most people in the world, including Darwin, believed in special creation and that all species were unchanged. George-Louis Leclerc was one of the first naturalist that questioned this belief. He proposed in his 44 volume natural history book that species evolve due to environmental influences. Erasmus Darwin was also a naturalist who also suggested that species evolved. Erasmus Darwin noted that some species have vestigial structures which are anatomical structures that have no apparent function in the species currently but would have been useful for the specie's ancestors. The thinking of these early 18th century naturalist helped to change the mindset and thinking of the early 19th century naturalist.

By the early 19th century, biogeography was ignited through the efforts of Alexander von Humboldt, Charles Lyell, and Charles Darwin. The 19th-century fascination with natural history engendered a fervor to be the first to collect rare specimens with the goal of doing so before they became extinct by other such collectors. Although the work of many 18th and 19th century naturalists were to inspire nature enthusiasts and conservation organizations, their writings, by modern standards, showed insensitivity towards conservation as they would kill hundreds of specimens for their collections.

Conservation Movement

The modern roots of conservation biology can be found in the late 18th century Enlightenment period particularly in England and Scotland. A number of thinkers, among them notably Lord Monboddo, described the importance of "preserving nature"; much of this early emphasis had its origins in Christian theology.

Scientific conservation principles were first practically applied to the forests of British India. The conservation ethic that began to evolve included three core principles: that human activity damaged the environment, that there was a civic duty to maintain the environment for future generations, and that scientific, empirically based methods should be applied to ensure this duty was carried out. Sir James Ranald Martin was prominent in promoting this ideology, publishing many medico-topographical reports that demonstrated the scale of damage wrought through large-scale deforestation and desiccation, and lobbying extensively for the institutionalization of forest conservation activities in British India through the establishment of Forest Departments.

The Madras Board of Revenue started local conservation efforts in 1842, headed by Alexander Gibson, a professional botanist who systematically adopted a forest conservation program based on scientific principles. This was the first case of state conservation management of forests in the world. Governor-General Lord Dalhousie introduced the first permanent and large-scale forest conservation program in the world in 1855, a model that soon spread to other colonies, as well the United States, where Yellowstone National Park was opened in 1872 as the world's first national park.

The term *conservation* came into widespread use in the late 19th century and referred to the management, mainly for economic reasons, of such natural resources as timber, fish, game, topsoil, pastureland, and minerals. In addition it referred to the preservation of forests (forestry), wildlife (wildlife refuge), parkland, wilderness, and watersheds. This period also saw the passage of the first conservation legislation and the establishment of the first nature conservation societies. The Sea Birds Preservation Act of 1869 was passed in Britain as the first nature protection law in the world after extensive lobbying from the Association for the Protection of Seabirds and the respected ornithologist Alfred Newton. Newton was also instrumental in the passage of the first Game laws from 1872, which protected animals during their breeding season so as to prevent the stock from being brought close to extinction.

One of the first conservation societies was the Royal Society for the Protection of Birds, founded in 1889 in Manchester as a protest group campaigning against the use of great crested grebe and kittiwake skins and feathers in fur clothing. Originally known as "the Plumage League".the group gained popularity and eventually amalgamated with the Fur and Feather League in Croydon, and formed the RSPB. The National Trust formed in 1895 with the manifesto to "...promote the permanent preservation, for the benefit of the nation, of lands, ...to preserve (so far practicable) their natural aspect."

In the United States, the Forest Reserve Act of 1891 gave the President power to set aside forest reserves from the land in the public domain. John Muir founded the Sierra Club in 1892, and the New York Zoological Society was set up in 1895. A series of national forests and preserves were established by Theodore Roosevelt from 1901 to 1909. The 1916 National Parks Act, included a 'use without impairment' clause, sought by John Muir, which eventually resulted to the removal of a proposal to build a dam in Dinosaur National Monument in 1959.

Roosevelt and Muir on Glacier Point in Yosemite National Park.

In the 20th century, Canadian civil servants, including Charles Gordon Hewitt and James Harkin spearheaded the movement toward wildlife conservation.

Global Conservation Efforts

In the mid-20th century, efforts arose to target individual species for conservation, notably efforts in big cat conservation in South America led by the New York Zoological Society. In the early 20th century, the New York Zoological Society was instrumental in developing concepts of establishing preserves for particular species and conducting the necessary conservation studies to determine the suitability of locations that are most appropriate as conservation priorities; the work of Henry Fairfield Osborn Jr., Carl E. Akeley, Archie Carr and Archie Carr III is notable in this era. Akeley for example, having led expeditions to the Virunga Mountains and observed the mountain gorilla in the wild, became convinced that the species and the area were conservation priorities. He was instrumental in persuading Albert I of Belgium to act in defense of the mountain gorilla and establish Albert National Park (since renamed Virunga National Park) in what is now Democratic Republic of Congo.

By the 1970s, hundreds of species-specific protection plan ensued. This was primarily led by the United States under the Endangered Species Act, along with the Species at RIsk Act (SARA) of Canada, and Biodiversity Action Plans developed in Australia, Sweden, and United Kingdom. Notably the United Nations acted to conserve sites of outstanding cultural or natural importance to the common heritage of mankind. The programme was adopted by the General Conference of UNESCO in 1972. As of 2006, a total of 830 sites are listed: 644 cultural, and 162 natural. The first country to pursue aggressive biological conservation through national legislation was the United States, which passed back to back legislation in the Endangered Species Act (1966) and National Environmental Policy Act (1970), which together injected major funding and protection measures to large-scale habitat protection and threatened species research. Other conservation developments, however, have taken hold throughout the world. India, for example, passed the Wildlife Protection Act of 1972.

In 1980, a significant development was the emergence of the urban conservation movement. A local organization was established in Birmingham, UK, A development followed in rapid succession in cities across the UK, then overseas. Although perceived as a grassroots movement, its early development was driven by academic research into urban wildlife. Initially perceived as radical, the movement's view of conservation being inextricably linked with other human activity has now become mainstream in conservation thought. Considerable research effort is now directed at urban conservation biology. The Society for Conservation Biology originated in 1985.

By 1992 most of the countries of the world had become committed to the principles of conservation of biological diversity through the Convention on Biological Diversity; subsequently many countries began programmes of Biodiversity Action Plans to identify and conserve threatened species within their borders, as well as protect associated habitats. The late 1990s saw increasing professionalism in the sector, with the maturing of organisations such as the Institute of Ecology and Environmental Management and the Society for the Environment.

Since 2000, the concept of landscape scale conservation has risen to prominence, with less emphasis being given to single-species or even single-habitat focused actions. Instead, an ecosystem approach is advocated by most mainstream conservationists, although concerns have been expressed by those working to protect some high-profile species.

Ecology has clarified the workings of the biosphere; i.e., the complex interrelationships among humans, other species, and the physical environment. The burgeoning human population and associated agriculture, industry, and the ensuing pollution, have demonstrated how easily ecological relationships can be disrupted.

CONCEPTS AND FOUNDATIONS

Measuring Extinction Rates

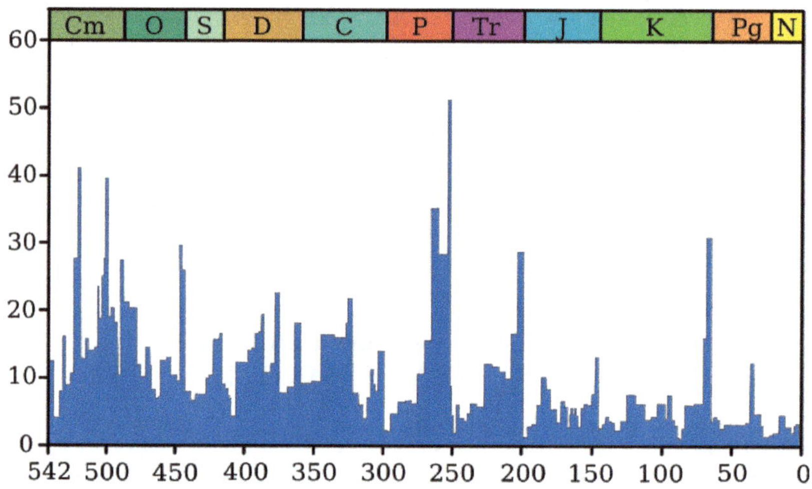

The blue graph shows the apparent percentage (not the absolute number) of marine animal genera becoming extinct during any given time interval. It does not represent all marine species, just those that are readily fossilized. The labels of the "Big Five" extinction events are clickable hyperlinks; see Extinction event for more details. (source and image info)

Extinction rates are measured in a variety of ways. Conservation biologists measure and apply statistical measures of fossil records, rates of habitat loss, and a multitude of other variables such as loss of biodiversity as a function of the rate of habitat loss and site occupancy to obtain such estimates. *The Theory of Island Biogeography* is possibly the most significant contribution toward the scientific understanding of both the process and how to measure the rate of species extinction. The current background extinction rate is estimated to be one species every few years.

The measure of ongoing species loss is made more complex by the fact that most of the Earth's species have not been described or evaluated. Estimates vary greatly on how many species actually exist (estimated range: 3,600,000-111,700,000) to how many have received a species binomial (estimated range: 1.5-8 million). Less than 1% of all species that have been described have been studied beyond simply noting its existence. From these figures, the IUCN reports that 23% of vertebrates, 5% of invertebrates, and 70% of plants that have been evaluated are designated as endangered or threatened. Better knowledge is being constructed by The Plant List for actual numbers of species.

Systematic Conservation Planning

Systematic conservation planning is an effective way to seek and identify efficient and effective types of reserve design to capture or sustain the highest priority biodiversity values and to work with communities in support of local ecosystems. Margules and

Pressey identified six interlinked stages in the systematic planning approach:

- Compile data on the biodiversity of the planning region
- Identify conservation goals for the planning region
- Review existing conservation areas
- Select additional conservation areas
- Implement conservation actions
- Maintain the required values of conservation areas

Conservation biologists regularly prepare detailed conservation plans for grant proposals or to effectively coordinate their plan of action and to identify best management practices (e.g. Systematic strategies generally employ the services of Geographic Information Systems (GIS) to assist in the decision-making process).

Conservation Physiology: a Mechanistic Approach to Conservation

Conservation Physiology was defined by Steven J. Cooke and colleagues as: 'An integrative scientific discipline applying physiological concepts, tools, and knowledge to characterizing biological diversity and its ecological implications; understanding and predicting how organisms, populations, and ecosystems respond to environmental change and stressors; and solving conservation problems across the broad range of taxa (i.e. including microbes, plants, and animals). Physiology is considered in the broadest possible terms to include functional and mechanistic responses at all scales, and conservation includes the development and refinement of strategies to rebuild populations, restore ecosystems, inform conservation policy, generate decision-support tools, and manage natural resources.' Conservation physiology is particularly relevant to practitioners in that it has the potential to generate cause-and-effect relationships and reveal the factors that contribute to population declines.

Conservation Biology as a Profession

The Society for Conservation Biology is a global community of conservation professionals dedicated to advancing the science and practice of conserving biodiversity. Conservation biology as a discipline reaches beyond biology, into subjects such as philosophy, law, economics, humanities, arts, anthropology, and education. Within biology, conservation genetics and evolution are immense fields unto themselves, but these disciplines are of prime importance to the practice and profession of conservation biology.

Is conservation biology an objective science when biologists advocate for an inherent value in nature? Do conservationists introduce bias when they support policies using qualitative description, such as habitat *degradation*, or *healthy* ecosystems? As all scientists hold values, so do conservation biologists. Conservation biologists advocate for reasoned and sensible management of natural resources and do so with a disclosed combination of science, reason, logic, and values in their conservation management plans. This sort of advocacy is similar to the medical profession advocating for healthy lifestyle options, both are beneficial to human well-being yet remain scientific in their approach.

There is a movement in conservation biology suggesting a new form of leadership is needed to mobilize conservation biology into a more effective discipline that is able to communicate the full scope of the problem to society at large. The movement proposes an adaptive leadership approach that parallels an adaptive management approach. The concept is based on a new philosophy or leadership theory steering away from historical notions of power, authority, and dominance. Adaptive conservation leadership is reflective and more equitable as it applies to any member of society who can mobilize others toward meaningful change using communication techniques that are inspiring, purposeful, and collegial. Adaptive conservation leadership and mentoring programs are being implemented by conservation biologists through organizations such as the Aldo Leopold Leadership Program

Approaches

Conservation may be classified as either in-situ conservation, which is protecting an endangered species in its natural habitat, or ex-situ conservation, which occurs outside the natural habitat. In-situ conservation involves protecting or restoring the habitat. Ex-situ conservation, on the other hand, involves protection outside of an organism's natural habitat, such as on reservations or in gene banks, in circumstances where viable populations may not be present in the natural habitat.

Also, non-interference may be used, which is termed a preservationist method. Preservationists advocate for giving areas of nature and species a protected existence that halts interference from the humans. In this regard, conservationists differ from preservationists in the social dimension, as conservation biology engages society and seeks equitable solutions for both society and ecosystems. Some preservationists emphasize the potential of biodiversity in a world without humans.

Ethics and Values

Conservation biologists are interdisciplinary researchers that practice ethics in the biological and social sciences. Chan states that conservationists must advocate for biodiversity and can do so in a scientifically ethical manner by not promoting simultaneous advocacy against other competing values. A conservationist researches biodiversity and reasons through a Resource Conservation Ethic , which identify what measures will deliver "the greatest good for the greatest number of people for the longest time."

Some conservation biologists argue that nature has an intrinsic value that is independent of anthropocentric usefulness or utilitarianism. Intrinsic value advocates that a gene, or species, be valued because they have a utility for the ecosystems they sustain. Aldo Leopold was a classical thinker and writer on such conservation ethics whose philosophy, ethics and writings are still valued and revisited by modern conservation biologists. His writing is often required reading for those in the profession.

Conservation Priorities

A pie chart image showing the relative biomass representation in a rain forest through a summary of children's perceptions from drawings and artwork (left), through a scientific estimate of actual biomass (middle), and by a measure of biodiversity (right). Notice that the biomass of social insects (middle) far outweighs the number of species (right).

The International Union for the Conservation of Nature (IUCN) has organized a global assortment of scientists and research stations across the planet to monitor the changing state of nature in an effort to tackle the extinction crisis. The IUCN provides annual updates on the status of species conservation through its Red List. The IUCN Red List serves as an international conservation tool to identify those species most in need of conservation attention and by providing a global index on the status of biodiversity. More than the dramatic rates of species loss, however, conservation scientists note that the sixth mass extinction is a biodiversity crisis requiring far more action than a priority focus on rare, endemic or endangered species. Concerns for biodiversity loss covers a broader conservation mandate that looks at ecological processes, such as migration, and a holistic examination of biodiversity at levels beyond the species, including genetic, population and ecosystem diversity. Extensive, systematic, and rapid rates of biodiversity loss threatens the sustained well-being of humanity by limiting supply of ecosystem services that are otherwise regenerated by the complex and evolving holistic network of genetic and ecosystem diversity. While the conservation status of species is employed extensively in conservation management, some scientists highlight that it is the common species that are the primary source of exploitation and habitat alteration by humanity. Moreover, common species are often undervalued despite their role as the primary source of ecosystem services.

While most in the community of conservation science "stress the importance" of sustaining biodiversity, there is debate on how to prioritize genes, species, or ecosystems, which are all components of biodiversity (e.g. Bowen, 1999). While the predominant approach to date has been to focus efforts on endangered species by conserving *biodiversity hotspots*, some scientists (e.g) and conservation organizations, such as the Nature Conservancy, argue that it is more cost-effective, logical, and socially relevant to invest in *biodiversity coldspots*. The costs of discovering, naming, and mapping out the distribution of every species, they argue, is an ill-advised conservation venture. They reason it is better to understand the significance of the ecological roles of species.

Biodiversity hotspots and coldspots are a way of recognizing that the spatial concentration of genes, species, and ecosystems is not uniformly distributed on the

Earth's surface. For example, "[...] 44% of all species of vascular plants and 35% of all species in four vertebrate groups are confined to 25 hotspots comprising only 1.4% of the land surface of the Earth."

Those arguing in favor of setting priorities for coldspots point out that there are other measures to consider beyond biodiversity. They point out that emphasizing hotspots downplays the importance of the social and ecological connections to vast areas of the Earth's ecosystems where biomass, not biodiversity, reigns supreme. It is estimated that 36% of the Earth's surface, encompassing 38.9% of the worlds vertebrates, lacks the endemic species to qualify as biodiversity hotspot. Moreover, measures show that maximizing protections for biodiversity does not capture ecosystem services any better than targeting randomly chosen regions. Population level biodiversity (i.e. coldspots) are disappearing at a rate that is ten times that at the species level. The level of importance in addressing biomass versus endemism as a concern for conservation biology is highlighted in literature measuring the level of threat to global ecosystem carbon stocks that do not necessarily reside in areas of endemism. A hotspot priority approach would not invest so heavily in places such as steppes, the Serengeti, the Arctic, or taiga. These areas contribute a great abundance of population (not species) level biodiversity and ecosystem services, including cultural value and planetary nutrient cycling.

Those in favor of the hotspot approach point out that species are irreplaceable components of the global ecosystem, they are concentrated in places that are most threatened, and should therefore receive maximal strategic protections. The IUCN Red List categories, which appear on Wikipedia species articles, is an example of the hotspot conservation approach in action; species that are not rare or endemic are listed the least concern and their Wikipedia articles tend to be ranked low on the importance scale. This is a hotspot approach because the priority is set to target species level concerns over population level or biomass. Species richness and genetic biodiversity contributes to and engenders ecosystem stability, ecosystem processes, evolutionary adaptability, and biomass. Both sides agree, however, that conserving biodiversity is necessary to reduce the extinction rate and identify an inherent value in nature; the debate hinges on how to prioritize limited conservation resources in the most cost-effective way.

Economic Values and Natural Capital

Conservation biologists have started to collaborate with leading global economists to determine how to measure the wealth and services of nature and to make these values apparent in global market transactions. This system of accounting is called *natural capital* and would, for example, register the value of an ecosystem before it is cleared to make way for development. The WWF publishes its *Living Planet Report* and provides a global index of biodiversity by monitoring approximately 5,000 populations in 1,686 species of vertebrate (mammals, birds, fish, reptiles, and amphibians) and report on the trends in much the same way that the stock market is tracked.

Tadrart Acacus desert in western Libya, part of the Sahara.

This method of measuring the global economic benefit of nature has been endorsed by the G8+5 leaders and the European Commission. Nature sustains many ecosystem services that benefit humanity. Many of the earths ecosystem services are public goods without a market and therefore no price or value. When the *stock market* registers a financial crisis, traders on Wall Street are not in the business of trading stocks for much of the planet's living natural capital stored in ecosystems. There is no natural stock market with investment portfolios into sea horses, amphibians, insects, and other creatures that provide a sustainable supply of ecosystem services that are valuable to society. The ecological footprint of society has exceeded the bio-regenerative capacity limits of the planet's ecosystems by about 30 percent, which is the same percentage of vertebrate populations that have registered decline from 1970 through 2005.

The inherent natural economy plays an essential role in sustaining humanity, including the regulation of global atmospheric chemistry, pollinating crops, pest control, cycling soil nutrients, purifying our water supply, supplying medicines and health benefits, and unquantifiable quality of life improvements. There is a relationship, a correlation, between markets and natural capital, and social income inequity and biodiversity loss. This means that there are greater rates of biodiversity loss in places where the inequity of wealth is greatest

Although a direct market comparison of natural capital is likely insufficient in terms of human value, one measure of ecosystem services suggests the contribution amounts to trillions of dollars yearly. For example, one segment of North American forests has been assigned an annual value of 250 billion dollars; as another example, honey-bee pollination is estimated to provide between 10 and 18 billion dollars of value yearly. The value of ecosystem services on one New Zealand island has been imputed to be as great as the GDP of that region. This planetary wealth is being lost at an incredible rate as the demands of human society is exceeding the bio-regenerative capacity of the Earth.

While biodiversity and ecosystems are resilient, the danger of losing them is that humans cannot recreate many ecosystem functions through technological innovation.

Strategic Species Concepts

Keystone Species

Some species, called a *keystone species* form a central supporting hub unique to their the ecosystem. The loss of such a species results in a collapse in ecosystem function, as well as the loss of coexisting species. Keystone species are usually predators due to their ability to control the population of prey in their ecosystem. The importance of a keystone species was shown by the extinction of the Steller's sea cow (*Hydrodamalis gigas*) through its interaction with sea otters, sea urchins, and kelp. Kelp beds grow and form nurseries in shallow waters to shelter creatures that support the food chain. Sea urchins feed on kelp, while sea otters feed on sea urchins. With the rapid decline of sea otters due to overhunting, sea urchin populations grazed unrestricted on the kelp beds and the ecosystem collapsed. Left unchecked, the urchins destroyed the shallow water kelp communities that supported the Steller's sea cow's diet and hastened their demise. The sea otter was thought to be a keystone species because the coexistence of many ecological associates in the kelp beds relied upon otters for their survival. However this was later questioned by Turvey and Risley, who showed that hunting alone would have driven the Steller's sea cow extinct.

Indicator Species

An *indicator species* has a narrow set of ecological requirements, therefore they become useful targets for observing the health of an ecosystem. Some animals, such as amphibians with their semi-permeable skin and linkages to wetlands, have an acute sensitivity to environmental harm and thus may serve as a *miner's canary*. Indicator species are monitored in an effort to capture environmental degradation through pollution or some other link to proximate human activities. Monitoring an indicator species is a measure to determine if there is a significant environmental impact that can serve to advise or modify practice, such as through different forest silviculture treatments and management scenarios, or to measure the degree of harm that a pesticide may impart on the health of an ecosystem.

Government regulators, consultants, or NGOs regularly monitor indicator species, however, there are limitations coupled with many practical considerations that must be followed for the approach to be effective. It is generally recommended that multiple indicators (genes, populations, species, communities, and landscape) be monitored for effective conservation measurement that prevents harm to the complex, and often unpredictable, response from ecosystem dynamics (Noss, 1997).

Umbrella and Flagship Species

An example of an *umbrella species* is the monarch butterfly, because of its lengthy migrations and aesthetic value. The monarch migrates across North America, covering multiple ecosystems and so requires a large area to exist. Any protections afforded to the monarch butterfly will at the same time umbrella many other species and habitats. An umbrella species is often used as *flagship species*, which are species, such as the giant panda, the blue whale, the tiger, the mountain gorilla and the monarch butterfly, that capture the public's attention and attract support for conservation measures.

Context and Trends

Conservation biologists study trends and process from the paleontological past to the ecological present as they gain an understanding of the context related to species extinction. It is generally accepted that there have been five major global mass extinctions that register in Earth's history. These include: the Ordovician (440 mya), Devonian (370 mya), Permian–Triassic (245 mya), Triassic–Jurassic (200 mya), and Cretaceous–Paleogene extinction event (66 mya) extinction spasms. Within the last 10,000 years, human influence over the Earth's ecosystems has been so extensive that scientists have difficulty estimating the number of species lost; that is to say the rates of deforestation, reef destruction, wetland draining and other human acts are proceeding much faster than human assessment of species. The latest *Living Planet Report* by the World Wide Fund for Nature estimates that we have exceeded the bio-regenerative capacity of the planet, requiring 1.5 Earths to support the demands placed on our natural resources.

Holocene Extinction

An art scape image showing the relative importance of animals in a rain forest through a summary of (a) child's perception compared with (b) a scientific estimate of the importance. The size of the animal represents its importance. The child's mental image places importance on big cats, birds, butterflies, and then reptiles versus the actual dominance of social insects (such as ants).

Conservation biologists are dealing with and have published evidence from all corners of the planet indicating that humanity may be causing the sixth and fastest planetary extinction event. It has been suggested that we are living in an era of unprecedented numbers of species extinctions, also known as the Holocene extinction event. The global extinction rate may be approximately 1,000 times higher than the natural background extinction rate. It is estimated that two-thirds of all mammal genera and one-half of all mammal species weighing at least 44 kilograms (97 lb) have gone extinct in the last 50,000 years. The Global Amphibian Assessment reports that amphibians are declining on a global scale faster than any other vertebrate group, with over 32% of all surviving species being threatened with extinction. The surviving populations are in continual decline in 43% of those that are threatened. Since the mid-1980s the actual rates of extinction have exceeded 211 times rates measured from the fossil record. However, "The current amphibian extinction rate may range from 25,039 to 45,474 times the background extinction rate for amphibians." The global extinction trend occurs in every major vertebrate group that is being monitored. For example, 23% of all mammals and 12% of all birds are Red Listed by the International Union for Conservation of Nature (IUCN), meaning they too are threatened with extinction. Even though extinction is natural, the decline in species is happening at such an incredible rate that evolution can simply not match, therefore, leading to the greatest continual mass extinction on Earth. Humans have dominated the planet and our high consumption of resources, along with the pollution generated is affecting the environments in which other species live. There are a wide variety of species that humans are working to protect such as the Hawaiian Crow and the Whooping Crane of Texas. People can also take action on preserving species by advocating and voting for policies that improve climate. The Earth's oceans especially require attention, as climate change has altered pH levels making it inhabitable for organisms with shells, that are dissolving as a result.

Status of Oceans and Reefs

Global assessments of coral reefs of the world continue to report drastic and rapid rates of decline. By 2000, 27% of the world's coral reef ecosystems had effectively collapsed. The largest period of decline occurred in a dramatic "bleaching" event in 1998, where approximately 16% of all the coral reefs in the world disappeared in less than a year. *Coral bleaching* is caused by a mixture of environmental stresses, including increases in ocean temperatures and acidity, causing both the release of symbiotic algae and death of corals. Decline and extinction risk in coral reef biodiversity has risen dramatically in the past ten years. The loss of coral reefs, which are predicted to go extinct in the next century, threatens the balance of global biodiversity, will have huge economic impacts, and endangers food security for hundreds of millions of people.Conservation biology plays an important role in international agreements covering the world's oceans (and other issues pertaining to biodiversity).

The oceans are threatened by acidification due to an increase in CO_2 levels. This is a most serious threat to societies relying heavily upon oceanic natural resources. A concern is that the majority of all marine species will not be able to evolve or acclimate in response to the changes in the ocean chemistry.

The prospects of averting mass extinction seems unlikely when "[...] 90% of all of the large (average approximately ≥50 kg), open ocean tuna, billfishes, and sharks in the ocean" are reportedly gone. Given the scientific review of current trends, the ocean is predicted to have few surviving multi-cellular organisms with only microbes left to dominate marine ecosystems.

Groups Other Than Vertebrates

Serious concerns also being raised about taxonomic groups that do not receive the same degree of social attention or attract funds as the vertebrates. These include fungal (including lichen-forming species), invertebrate (particularly insect) and plant communities where the vast majority of biodiversity is represented. Conservation of fungi and conservation of insects, in particular, are both of pivotal importance for conservation biology. As mycorrhizal symbionts, and as decomposers and recyclers, fungi are essential for sustainability of forests. The value of insects in the biosphere is enormous because they outnumber all other living groups in measure of species richness. The greatest bulk of biomass on land is found in plants, which is sustained by insect relations. This great ecological value of insects is countered by a society that often reacts negatively toward these aesthetically 'unpleasant' creatures.

One area of concern in the insect world that has caught the public eye is the mysterious case of missing honey bees (*Apis mellifera*). Honey bees provide an indispensable ecological services through their acts of pollination supporting a huge variety of agriculture crops. The use of honey and wax have become vastly used throughout the world. The sudden disappearance of bees leaving empty hives or colony collapse disorder (CCD) is not uncommon. However, in 16-month period from 2006 through 2007, 29% of 577 beekeepers across the United States reported CCD losses in up to 76% of their colonies. This sudden demographic loss in bee numbers is placing a strain on the agricultural sector. The cause behind the massive declines is puzzling scientists. Pests, pesticides, and global warming are all being considered as possible causes.

Another highlight that links conservation biology to insects, forests, and climate change is the mountain pine beetle (*Dendroctonus ponderosae*) epidemic of British Columbia, Canada, which has infested 470,000 km^2 (180,000 sq mi) of forested land since 1999. An action plan has been prepared by the Government of British Columbia to address this problem.

This impact converted the forest from a small net carbon sink to a large net carbon source both during and immediately after the outbreak. In the worst year, the impacts resulting from the beetle outbreak in British Columbia were equivalent to 75% of the average annual direct forest fire emissions from all of Canada during 1959–1999.

—Kurz et al.

Conservation Biology of Parasites

A large proportion of parasite species are threatened by extinction. A few of them are being eradicated as pests of humans or domestic animals, however, most of them are harmless. Threats include the decline or fragmentation of host populations, or the extinction of host species.

Threats to Biodiversity

Today, many threats to Biodiversity exist. An acronym that can be used to express the top threats of present-day H.I.P.P.O stands for Habitat Loss, Invasive Species, Pollution, Human Population, and Overharvesting. The primary threats to biodiversity are habitat destruction (such as deforestation, agricultural expansion, urban development), and overexploitation (such as wildlife trade). Habitat fragmentation also poses challenges, because the global network of protected areas only covers 11.5% of the Earth's surface. A significant consequence of fragmentation and lack of linked protected areas is the reduction of animal migration on a global scale. Considering that billions of tonnes of biomass are responsible for nutrient cycling across the earth, the reduction of migration is a serious matter for conservation biology.

However, human activities need not necessarily cause irreparable harm to the biosphere. With conservation management and planning for biodiversity at all levels, from genes to ecosystems, there are examples where humans mutually coexist in a sustainable way with nature. Even with the current threats to biodiversity there are ways we can improve the current condition and start anew.

Many of the threats to biodiversity, including disease and climate change, are reaching inside borders of protected areas, leaving them ‹not-so protected' (e.g. Yellowstone National Park). Climate change, for example, is often cited as a serious threat in this regard, because there is a feedback loop between species extinction and the release of carbon dioxide into the atmosphere. Ecosystems store and cycle large amounts of carbon which regulates global conditions. In present day, there have been major climate shifts with temperature changes making survival of some species difficult. The effects of global warming add a catastrophic threat toward a mass extinction of global biological diversity. Conservationists have claimed that not all the species can be saved, and they have to decide which their efforts should be used to protect. This concept is known as the Conservation Triage. The extinction threat is estimated to range from 15 to 37 percent of all species by 2050, or 50 percent of all species over the next 50 years. The current extinction rate is 100-100,000 times more rapid today than the last several billion years.

REFERENCES

1. Sahney, S.; Benton, M. J (2008). "Recovery from the most profound mass extinction of all time". Proceedings of the Royal Society B: Biological Sciences. 275 (1636): 759–65. doi:10.1098/rspb.2007.1370. PMC 2596898 . PMID 18198148.
2. Soulé, Michael E.; Wilcox, Bruce A. (1980). Conservation biology: an evolutionary-ecological perspective. Sunderland, Mass: Sinauer Associates. ISBN 0-87893-800-1.
3. Soulé (1986). "What is Conservation Biology?" (PDF). BioScience. American Institute of Biological Sciences. 35 (11): 727–34. doi:10.2307/1310054. JSTOR

1310054. |first2= missing |last2=in Authors list (help)

4. Soule, Michael E. (1986). Conservation Biology: The Science of Scarcity and Diversity. Sinauer Associates. p. 584. ISBN 978-0-87893-795-0.
5. Hunter, Malcolm L. (1996). Fundamentals of conservation biology. Oxford: Blackwell Science. ISBN 0-86542-371-7.
6. Meffe, Gary K.; Martha J. Groom (2006). Principles of conservation biology (3rd ed.). Sunderland, Mass: Sinauer Associates. ISBN 0-87893-518-5.
7. van Dyke, Fred (2008). Conservation Biology: Foundations, Concepts, Applications (2nd ed.). Springer Verlag. p. 478. ISBN 978-1-4020-6890-4.
8. J. Douglas. 1978. Biologists urge US endowment for conservation. Nature Vol. 275, 14 September 1978. Kat Williams . 1978. Natural Sciences. Science News. September 30, 1978.
9. Organization of the meeting itself also entailed bridging a gap between genetics and ecology. Soulé, was an evolutionary geneticist working with wheat geneticist Sir Otto Frankel to advance conservation genetics as a new field at the time. Jared Diamond, who suggested the idea for a conference to Wilcox was concerned with the application of community ecology and island biogeography theory to conservation. Wilcox and Thomas Lovejoy, who together initiated planning for the conference in June 1977 when Lovejoy secured a commitment of seed funding at World Wildlife Fund, felt both genetics and ecology should be represented. Wilcox suggested use of a new term conservation biology, complementing Frankel's conception and coining of "conservation genetics", to encompass the application of biological sciences in general to conservation. Subsequently, Soulé and Wilcox wrote conceived the agenda for the meeting they jointly convened on September 6–9, 1978, titled First International Conference on Resesarch in Conservation Biology, in which the program described "The purpose of this conference is to accelerate and facilitate the development of a rigorous new discipline called conservation biology -- a multidisciplinary field drawing its insights and methodology mostly from population ecology, community ecology, sociobiology, population genetics, and reproductive biology." This inclusion of topics at the meeting related to animal breeding reflected participation and support of the zoo and captive breeding communities.
10. Cooke, S. J.; Sack, L.; Franklin, C. E.; Farrell, A. P.; Beardall, J.; Wikelski, M.; Chown, S. L. (2013). "What is conservation physiology? Perspectives on an increasingly integrated and essential science". Conservation Physiology. 1 (1): cot001. doi:10.1093/conphys/cot001. PMC 4732437 . PMID 27293585.
11. Wilson, Edward Osborne (2002). The future of life. Boston: Little, Brown. ISBN 0-316-64853-1.
12. Kala, Chandra Prakash (2005). "Indigenous Uses, Population Density, and Conservation of Threatened Medicinal Plants in Protected Areas of the Indian Himalayas". Conservation Biology. 19 (2): 368–78. doi:10.1111/j.1523-1739.2005.00602.x. JSTOR 3591249.
13. Sahney, S.; Benton, M. J.; Ferry, P. A. (2010). "Links between global taxonomic

diversity, ecological diversity and the expansion of vertebrates on land". Biology Letters. 6 (4): 544–7. doi:10.1098/rsbl.2009.1024. PMC 2936204 . PMID 20106856.

14. Koh, Lian Pin; Dunn, Robert R.; Sodhi, Navjot S.; Colwell, Robert K.; Proctor, Heather C.; Smith, Vincent S. (2004). "Species coextinctions and the biodiversity crisis". Science. 305 (5690): 1632–4. Bibcode:2004Sci...305.1632K. doi:10.1126/science.1101101. PMID 15361627.
15. Millennium Ecosystem Assessment (2005). Ecosystems and Human Well-being: Biodiversity Synthesis. World Resources Institute, Washington, DC.
16. Jackson, J. B. C. (2008). "Ecological extinction and evolution in the brave new ocean". Proceedings of the National Academy of Sciences. 105 (Suppl 1): 11458–65. Bibcode:2008PNAS..10511458J. doi:10.1073/pnas.0802812105. PMC 2556419 . PMID 18695220.
17. Theodore Roosevelt, Address to the Deep Waterway Convention Memphis, TN, October 4, 1907
18. ref_churchill. "Biodiversity protection and preservation". www.ffem.fr. Retrieved 2016-10-11.
19. Hardin G (December 1968). "The Tragedy of the Commons" (PDF). Science. 162 (3859): 1243–8. Bibcode:1968Sci...162.1243H. doi:10.1126/science.162.3859.1243. PMID 5699198. Archived from the original (PDF) on 2011-07-19.
20. Also considered to be a consequence of evolution, where individual selection is favored over group selection. For recent discussions, see: Kay CE (1997). "The Ultimate Tragedy of Commons". Conserv. Biol. 11 (6): 1447–8. doi:10.1046/j.1523-1739.1997.97069.x.
21. and Wilson DS, Wilson EO (December 2007). "Rethinking the theoretical foundation of sociobiology" (PDF). Q Rev Biol. 82 (4): 327–48. doi:10.1086/522809. PMID 18217526. Archived from the original on 2009-03-26.
22. Mason, Rachel and Judith Ramos. (2004). Traditional Ecological Knowledge of Tlingit People concerning the Sockeye Salmon Fishery of the Dry Bay Area, A Cooperative Agreement Between Department of the Interior National Park Service and the Yakutat Tlingit Tribe, Final Report (FIS) Project 01-091, Yakutat, Alaska.
23. Murphree, Marshall W. (2009-05-22). "The strategic pillars of communal natural resource management: benefit, empowerment and conservation". Biodiversity and Conservation. 18 (10): 2551–2562. doi:10.1007/s10531-009-9644-0. ISSN 0960-3115.
24. Wilson, David Alec (2002). Darwin's cathedral: evolution, religion, and the nature of society. Chicago: University of Chicago Press. ISBN 0-226-90134-3.
25. Primack, Richard B. (2004). A Primer of Conservation Biology, 3rd ed. Sinauer Associates. pp. 320pp. ISBN 0-87893-728-5.
26. Hamilton, E., and H. Cairns (eds). 1961. Plato: the collected dialogues. Princeton University Press, Princeton, NJ

27. The Bible, Leviticus, 25:4-5
28. Evans, David (1997). A history of nature conservation in Britain. New York: Routledge. ISBN 0-415-14491-4.
29. Farber, Paul Lawrence (2000). Finding order in nature: the naturalist tradition from Linnaeus to E. O. Wilson. Baltimore: Johns Hopkins University Press. ISBN 0-8018-6390-2.
30. Mader, Sylvia (2016). Biology. New York, NY: McGraw Hill Education. p. 262. ISBN 978-0-07-802426-9.
31. Short history of biogeography and conservation biology
32. Cloyd, E. L. (1972). James Burnett, Lord Monboddo. New York: Oxford University Press. p. 196. ISBN 0-19-812437-6.
33. Stebbing, E.P (1922)The forests of India vol. 1, pp. 72-81
34. Barton, Greg (2002). Empire Forestry and the Origins of Environmentalism. Cambridge University Press. p. 48. ISBN 978-1-139-43460-7.
35. MUTHIAH, S. (Nov 5, 2007). "A life for forestry". Metro Plus Chennai. Chennai, India: The Hindu. Retrieved 2009-03-09.
36. Cleghorn, Hugh Francis Clarke (1861). The Forests and Gardens of South India (Original from the University of Michigan, Digitized Feb 10, 2006 ed.). London: W. H. Allen. OCLC 301345427.
37. America has been the context for both the origins of conservation history and its modern form, environmental history. Asiaticsociety.org.bd. Retrieved on 2011-09-01.
38. Haines, Aubrey (1996). The Yellowstone Story: A History of Our First National Park: Volume 1 Revised Edition. Yellowstone Association for Natural Science, History of Education.
39. G. Baeyens; M. L. Martinez (2007). Coastal Dunes: Ecology and Conservation. Springer. p. 282.
40. Makel, Jo (2 February 2011). "Protecting seabirds at Bempton Cliffs". BBC News.
41. Newton A. 1899. The plume trade: borrowed plumes. The Times 28 January 1876; and The plume trade. The Times 25 February 1899. Reprinted together by the Society for thc Protcction of Birds, April 1899.
42. Newton A. 1868. The zoological aspect of game laws. Address to the British Association, Section D, August 1868. Reprinted [n.d.] by the Society for the Protection of Birds.
43. "Milestones". RSPB. Retrieved 19 February 2007.
44. Penna (1999), p. 99.
45. "History of the RSPB". RSPB. Retrieved 19 February 2007.
46. "Theodore Roosevelt and Conservation - Theodore Roosevelt National Park (U.S. National Park Service)". www.nps.gov. Retrieved 2016-10-04.
47. Environmental timeline 1890–1920

48. Davis, Peter (1996). Museums and the natural environment: the role of natural history museums in biological conservation. London: Leicester University Press. ISBN 0-7185-1548-X.
49. For a review and introduction to the history of wildlife conservation and preservation in Canada, see Foster, Janet (1997). Working for wildlife: the beginning of preservation in Canada (2nd ed.). Toronto: University of Toronto Press. ISBN 0-8020-7969-5.
50. A.R. Rabinowitz, Jaguar: One Man's Battle to Establish the World's First Jaguar Preserve, Arbor House, New York, N.Y. (1986)
51. Carr, Marjorie Harris; Carr, Archie Fairly (1994). A naturalist in Florida: a celebration of Eden. New Haven, Conn: Yale University Press. ISBN 0-300-05589-7.
52. Henry Fairfield Osborn Jr. Biographical summary
53. Akeley, C., 1923. In Brightest Africa New York, Doubleday. 188-249.
54. U.S. Endangered Species Act (7 U.S.C. § 136, 16 U.S.C. § 1531 et seq.) of 1973, Washington DC, U.S. Government Printing Office
55. U.S. Endangered Species Act of 1966 with subsequent amendments
56. 42 USC 4321 National Environmental Policy Act (2000): full text of the law
57. Krausman, Paul R.; Johnsingh, A. J. T. (1990). "Conservation and Wildlife Education in India". Wildlife Society Bulletin. 18 (3): 342–7. JSTOR 3782224.
58. Dyke, Fred Van (2008-04-07). "Conservation biology: Foundations, concepts, applications". ISBN 978-1-4020-6890-4.
59. Convention on Biological Diversity Official Page
60. Gore, Albert (1992). Earth in the balance: ecology and the human spirit. Boston: Houghton Mifflin. ISBN 0-395-57821-3.
61. Regan, Helen M.; Lupia, Richard; Drinnan, Andrew N.; Burgman, Mark A. (2001). "The Currency and Tempo of Extinction". The American Naturalist. 157 (1): 1–10. doi:10.1086/317005. PMID 18707231.
62. MacKenzie, Darryl I.; Nichols, James D.; Hines, James E.; Knutson, Melinda G.; Franklin, Alan B. (2003). "Estimating Site Occupancy, Colonization, and Local Extinction When a Species Is Detected Imperfectly". Ecology. 84 (8): 2200–2207. doi:10.1890/02-3090. JSTOR 3450043.
63. Balmford, Andrew; Green, Rhys E.; Jenkins, Martin (2003). "Measuring the changing state of nature". Trends in Ecology & Evolution. 18 (7): 326–30. doi:10.1016/S0169-5347(03)00067-3.
64. MacArthur, R.H.; Wilson, E.O. (2001). The Theory of Island Biogeography. Princeton, N.J: Princeton University Press. ISBN 0-691-08836-5.
65. Raup DM (1991). "A kill curve for Phanerozoic marine species". Paleobiology. 17 (1): 37–48. PMID 11538288.
66. Wilson, Edward O. (2000). "On the Future of Conservation Biology". Conservation Biology. 14 (1): 1–3. doi:10.1046/j.1523-1739.2000.00000-e1.x.

67. IUCN Red-list statistics (2006) Archived June 30, 2006, at the Wayback Machine.

68. The IUCN does not disaggregate endangered from critically endangered or threatened for the purpose of these statistics.

69. Margules CR, Pressey RL (May 2000). "Systematic conservation planning" (PDF). Nature. 405 (6783): 243–53. doi:10.1038/35012251. PMID 10821285.

70. The Amphibian Conservation Action Plan is one example .

71. See also: Chan KM, Shaw MR, Cameron DR, Underwood EC, Daily GC (October 2006). "Conservation Planning for Ecosystem Services". PLoS Biol. 4 (11): e379. doi:10.1371/journal.pbio.0040379. PMC 1629036 . PMID 17076586.

72. Manolis JC, Chan KM, Finkelstein ME, Stephens S, Nelson CR, Grant JB, Dombeck MP (2009). "Leadership: A new frontier in conservation science". Conserv. Biol.

73. http://leopoldleadership.org/content/[full citation needed]

74. Kala, Chandra Prakash (2009). "Medicinal plants conservation and enterprise development". Medicinal Plants - International Journal of Phytomedicines and Related Industries. 1 (2): 79–95. doi:10.5958/j.0975-4261.1.2.011.

75. Chan, Kai M. A. (2008). "Value and Advocacy in Conservation Biology: Crisis Discipline or Discipline in Crisis?". Conservation Biology. 22 (1): 1–3. doi:10.1111/j.1523-1739.2007.00869.x. PMID 18254846.

76. "Archived copy". Archived from the original on 2014-06-27. Retrieved 2013-10-20.

77. Vié, J. C.; Hilton-Taylor, C.; Stuart, S.N. (2009). "Wildlife in a Changing World – An Analysis of the 2008 IUCN Red List of Threatened Species" (PDF). Gland, Switzerland: IUCN: 180. Retrieved December 24, 2010"."

78. Molnar, J.; Marvier, M.; Kareiva, P. (2004). "The sum is greater than the parts" (PDF). Conservation Biology. 18 (6): 1670–1. doi:10.1111/j.1523-1739.2004.00l07.x.

79. Gaston, K.J. (2010). "Valuing common species". Science. 327 (5962): 154–155. Bibcode:2010Sci...327..154G. doi:10.1126/science.1182818. PMID 20056880.

80. Kearns, Carol Ann (2010). "Conservation of Biodiversity". Nature Education Knowledge. 3 (10): 7.

81. Committee on Recently Extinct Organisms. "Why Care About Species That Have Gone Extinct?". URL accessed July 30, 2006.

82. Luck, Gary W.; Daily, Gretchen C.; Ehrlich, Paul R. (2003). "Population diversity and ecosystem services". Trends in Ecology & Evolution. 18 (7): 331–6. doi:10.1016/S0169-5347(03)00100-9.

83. Turner, Will R.; Brandon, Katrina; Brooks, Thomas M.; Costanza, Robert; da Fonesca, Gustavo A. B.; Portela, Rosimeiry (2007). "Global Conservation of Biodiversity and Ecosystem Services". BioScience. 57 (10): 868–73. doi:10.1641/B571009.

84. Molnar J, Marvier M, Karieva P (2004). "The sum is greater than the parts".

Conserv. Biol. 18 (6): 1670–1. doi:10.1111/j.1523-1739.2004.00107.x.

85. Myers, Norman; Mittermeier, Russell A.; Mittermeier, Cristina G.; da Fonseca, Gustavo A. B.; Kent, Jennifer (2000). "Biodiversity hotspots for conservation priorities". Nature. 403 (6772): 853–8. Bibcode:2000Natur.403..853M. doi:10.1038/35002501. PMID 10706275.
86. Underwood EC, Shaw MR, Wilson KA; et al. (2008). Somers, Michael, ed. "Protecting Biodiversity when Money Matters: Maximizing Return on Investment". PLoS ONE. 3 (1): e1515. Bibcode:2008PLoSO...3.1515U. doi:10.1371/journal.pone.0001515. PMC 2212107 . PMID 18231601.
87. Leroux SJ, Schmiegelow FK (February 2007). "Biodiversity concordance and the importance of endemism". Conserv. Biol. 21 (1): 266–8; discussion 269–70. doi:10.1111/j.1523-1739.2006.00628.x. PMID 17298533.
88. Naidoo R, Balmford A, Costanza R, et al. (July 2008). "Global mapping of ecosystem services and conservation priorities". Proc. Natl. Acad. Sci. U.S.A. 105 (28): 9495–500. Bibcode:2008PNAS..105.9495N. doi:10.1073/pnas.0707823105. PMC 2474481 . PMID 18621701.
89. Wood CC, Gross MR (February 2008). "Elemental conservation units: communicating extinction risk without dictating targets for protection" (PDF). Conserv. Biol. 22 (1): 36–47. doi:10.1111/j.1523-1739.2007.00856.x. PMID 18254851.
90. Running, S. W. (2008). "Climate Change: Ecosystem Disturbance, Carbon, and Climate". Science. 321 (5889): 652–3. doi:10.1126/science.1159607. PMID 18669853.
91. Kurz, W. A.; Dymond, C. C.; Stinson, G.; Rampley, G. J.; Neilson, E. T.; Carroll, A. L.; Ebata, T.; Safranyik, L. (2008). "Mountain pine beetle and forest carbon feedback to climate change". Nature. 452 (7190): 987–90. Bibcode:2008Natur.452..987K. doi:10.1038/nature06777. PMID 18432244.
92. The Global Conservation Fund is an example of funding organization that excludes biodiversity coldspots in its strategic campaign.
93. Kareiva, Peter; Marvier, Michelle (2003). "Conserving Biodiversity Coldspots". American Scientist. 91 (4): 344–51. doi:10.1511/2003.4.344.
94. "The Biodiversity Hotspots".
95. The following papers are examples of research showing the relationship between biodiversity, biomass, and ecosystem stability:
96. Cardinale BJ, Wright JP, Cadotte MW, et al. (November 2007). "Impacts of plant diversity on biomass production increase through time because of species complementarity". Proc. Natl. Acad. Sci. U.S.A. 104 (46): 18123–8. Bibcode:2007PNAS..10418123C. doi:10.1073/pnas.0709069104. PMC 2084307 . PMID 17991772.
97. European Communities (2008). The economics of ecosystems and biodiversity. Interim Report (PDF). Wesseling, Germany: Welzel+Hardt. ISBN 978-92-79-08960-2.

98. R. Costanza, R. d'Arge, R. de Groot, S. Farberk, M. Grasso, B. Hannon, K. Limburg, S. Naeem, R. V. O'Neill, J. Paruelo, R. G. Raskin, P. Suttonkk and M. van den Belt. The value of the world's ecosystem services and natural capital" Nature 387: 253-260 "Archived copy"(PDF). Archived from the original (PDF) on 2009-12-26. Retrieved 2009-12-30.

99. WWF. "World Wildlife Fund" (pdf). Retrieved January 8, 2009.

100. "From the Ecological Society of America (ESA)".

101. Millennium Ecosystem Assessment. (2005). Ecosystems and Human Well-being: Biodiversity Synthesis. World Resources Institute, Washington, DC.

102. http://www.millenniumassessment.org/

103. Black, Richard (2008-12-22). "Bees get plants' pests in a flap". BBC News. Retrieved 2010-04-01.

104. Hermoso, Virgilio; Abell, R; Linke, S; Boon, P (2016). "The role of protected areas for freshwater biodiversity conservation: challenges and opportunities in a rapidly changing world.". Aquatic conservation - Marine and freshwater ecosystems. 26 (S1): 3–11. doi:10.1002/aqc.2681.

105. Mitchell R, Popham F (November 2008). "Effect of exposure to natural environment on health inequalities: an observational population study". Lancet. 372 (9650): 1655–60. doi:10.1016/S0140-6736(08)61689-X. PMID 18994663.

106. Mikkelson GM, Gonzalez A, Peterson GD (2007). Chave, Jerome, ed. "Economic Inequality Predicts Biodiversity Loss". PLoS ONE. 2 (5): e444. Bibcode:2007PLoSO...2..444M. doi:10.1371/journal.pone.0000444. PMC 1864998 . PMID 17505535.

107. Staff of World Resources Program. (1998). Valuing Ecosystem Services. World Resources 1998-99.

108. Committee on Noneconomic and Economic Value of Biodiversity, Board on Biology, Commission on Life Sciences, National Research Council. (1999). Perspectives on biodiversity: valuing its role in an everchanging world. Washington, D.C: National Academy Press. ISBN 0-309-06581-X.

109. Valuation of Ecosystem services : A Backgrounder

110. Ecosystem Services: Estimated value in trillions

111. Canadian Forest Congress: Carbon capture, water filtration, other boreal forest ecoservices worth estimated $250 billion/year

112. APIS, Volume 10, Number 11, November 1992, M.T. Sanford: Estimated value of honey bee pollination

113. Regional council, Waikato: The hidden economy

114. Society, National Geographic (2011-01-21). "keystone species". National Geographic Society. Retrieved 2016-10-11.

115. P. K. Anderson. (1996). Competition, predation, and the evolution and extinction of Steller's Sea Cow, Hydrodamalis gigas. Marine Mammal Science, 11(3):391-394

116. Turvey, S.T; Risley, C.L (2006). "Modelling the extinction of Steller's sea

cow". Biology Letters. 2 (1): 94–7. doi:10.1098/rsbl.2005.0415. PMC 1617197 . PMID 17148336.

117. Landres PB, Verner J, Thomas JW (1988). "Ecological Uses of Vertebrate Indicator Species: A Critique" (PDF). Conserv. Biol. 2 (4): 316–28. doi:10.1111/j.1523-1739.1988.tb00195.x.
118. Carroll, C. Dennis; Meffe, Gary K. (1997). Principles of conservation biology. Sunderland, Mass: Sinauer. ISBN 0-87893-521-5.
119. Ehrlich, Anne H.; Ehrlich, Paul R. (1981). Extinction: the causes and consequences of the disappearance of species. New York: Random House. ISBN 0-394-51312-6.
120. http://www.panda.org/news_facts/publications/living_planet_report/[full citation needed]
121. Wake, D. B.; Vredenburg, V. T. (2008). "Are we in the midst of the sixth mass extinction? A view from the world of amphibians". Proceedings of the National Academy of Sciences. 105 (Suppl 1): 11466–73. Bibcode:2008PNAS..10511466W. doi:10.1073/pnas.0801921105. PMC 2556420 . PMID 18695221.
122. http://www.millenniumassessment.org[full citation needed]
123. National Survey Reveals Biodiversity Crisis - Scientific Experts Believe we are in the Midst of Fastest Mass Extinction in Earth's History a "Biofact" from the American Natural History Museum's website
124. May, Robert Lewis; Lawton, John (1995). Extinction rates. Oxford [Oxfordshire]: Oxford University Press. ISBN 0-19-854829-X.
125. Dell'Amore, Christine (30 May 2014). "Species Extinction Happening 1,000 Times Faster Because of Humans?". National Geographic. National Geographic. Retrieved 11 October 2016.
126. Avise, J. C.; Hubbell, S. P.; Ayala, F. J. (2008). "In the light of evolution II: Biodiversity and extinction". Proceedings of the National Academy of Sciences. 105 Suppl 1: 11453–7. Bibcode:2008PNAS..10511453A. doi:10.1073/pnas.0802504105. PMC 2556414 . PMID 18695213.
127. Bentley, Molly (January 2, 2009). "Diamond clues to beasts' demise". BBC News.
128. Kennett, D. J.; Kennett, J. P.; West, A.; Mercer, C.; Hee, S. S. Q.; Bement, L.; Bunch, T. E.; Sellers, M.; Wolbach, W. S. (2009). "Nanodiamonds in the Younger Dryas Boundary Sediment Layer". Science. 323 (5910): 94. Bibcode:2009Sci...323...94K. doi:10.1126/science.1162819. PMID 19119227.
129. "Archived copy". Archived from the original on 2009-07-06. Retrieved 2009-10-11.[full citation needed]
130. McCallum, Malcolm L. (2007). "Amphibian Decline or Extinction? Current Declines Dwarf Background Extinction Rate". Journal of Herpetology. 41 (3): 483–91. doi:10.1670/0022-1511(2007)41[483:ADOECD]2.0.CO;2. JSTOR 4498614.
131. Vince, Gaia. "A looming mass extinction caused by humans". www.bbc.com. BBC. Retrieved 11 October 2016.
132. Tate, Karl. "The New Dying: How Human-Caused Extinction Affects the Planet

(Infographic)". Live Science. Live Science. Retrieved 11 October 2016.

133. Worrall, Simon (20 August 2016). "How the Current Mass Extinction of Animals Threatens Humans". National Geographic. National Geographic. Retrieved 11 October 2016.

134. Australian State of the Environment Committee. (2001). Australia state of the environment 2001: independent report to the Commonwealth Minister for the Environment and Heritage (PDF). Collingwood, VIC, Australia: CSIRO Publishing. ISBN 0-643-06745-0.

135. Carpenter, K. E.; Abrar, M.; Aeby, G.; Aronson, R. B.; Banks, S.; Bruckner, A.; Chiriboga, A.; Cortes, J.; Delbeek, J. C.; DeVantier, L.; Edgar, G. J.; Edwards, A. J.; Fenner, D.; Guzman, H. M.; Hoeksema, B. W.; Hodgson, G.; Johan, O.; Licuanan, W. Y.; Livingstone, S. R.; Lovell, E. R.; Moore, J. A.; Obura, D. O.; Ochavillo, D.; Polidoro, B. A.; Precht, W. F.; Quibilan, M. C.; Reboton, C.; Richards, Z. T.; Rogers, A. D.; Sanciangco, J.; Sheppard, A.; Sheppard, C.; Smith, J.; Stuart, S.; Turak, E.; Veron, J. E. N.; Wallace, C.; Weil, E.; Wood, E. (2008). "One-Third of Reef-Building Corals Face Elevated Extinction Risk from Climate Change and Local Impacts". Science. 321 (5888): 560–3. Bibcode:2008Sci...321..560C. doi:10.1126/science.1159196. PMID 18653892.

136. http://laws.justice.gc.ca/en/showdoc/cs/M-7.01///en?page=1[full citation needed]

137. The Royal Society. 2005. Ocean acidification due to increasing atmospheric carbon dioxide. Policy document 12/05. ISBN 0-85403-617-2 Download

138. "Orphans of Rio" (PDF). fungal-conservation.org. Retrieved 2011-07-09.

139. Thomas, JA; Telfer, MG; Roy, DB; Preston, CD; Greenwood, JJ; Asher, J; Fox, R; Clarke, RT; Lawton, JH (2004). "Comparative Losses of British Butterflies, Birds, and Plants and the Global Extinction Crisis". Science. 303 (5665): 1879–81. doi:10.1126/science.1095046. PMID 15031508.

140. Dunn, Robert R. (2005). "Modern Insect Extinctions, the Neglected Majority". Conservation Biology. 19 (4): 1030–6. doi:10.1111/j.1523-1739.2005.00078.x.

141. Wilson, Edward O. (1987). "The Little Things That Run the world (The Importance and Conservation of Invertebrates)". Conservation Biology. 1 (4): 344–6. doi:10.1111/j.1523-1739.1987.tb00055.x. JSTOR 2386020.

142. Samways, Michael J. (1993). "Insects in biodiversity conservation: some perspectives and directives". Biodiversity and Conservation. 2 (3): 258–82. doi:10.1007/BF00056672.

143. Society, National Geographic. "Honeybee." National Geographic. National Geographic, n.d. Web. 11 Oct. 2016.

144. Holden, C. (2006). "Ecology: Report Warns of Looming Pollination Crisis in North America". Science. 314 (5798): 397. doi:10.1126/science.314.5798.397. PMID 17053115.

145. Stokstad, E. (2007). "Entomology: The Case of the Empty Hives". Science. 316 (5827): 970–2. doi:10.1126/science.316.5827.970. PMID 17510336.

146. http://www.for.gov.bc.ca/hfp/mountain_pine_beetle/actionplan/2006/Beetle_

Action_Plan.pdf[full citation needed]

147. "Threats to Biodiversity | GEOG 030: Geographic Perspectives on Sustainability and Human-Environment Systems, 2011". www.e-education.psu.edu. Retrieved 2016-10-07.

148. Freckleton, Rob; Sodhi, Navjot S.; Bickford, David; Diesmos, Arvin C.; Lee, Tien Ming; Koh, Lian Pin; Brook, Barry W.; Sekercioglu, Cagan H.; Bradshaw, Corey J. A. (2008). "Measuring the Meltdown: Drivers of Global Amphibian Extinction and Decline". PLoS ONE. 3 (2): e1636. Bibcode:2008PLoSO...3.1636S. doi:10.1371/journal.pone.0001636. PMC 2238793 . PMID 18286193.

149. Longcore, Travis; Rich, Catherine (2004). "Ecological light pollution". Frontiers in Ecology and the Environment. 2 (4): 191–8. doi:10.1890/1540-9295(2004)002[0191:ELP]2.0.CO;2. JSTOR 3868314.

150. "Asia's biodiversity vanishing into the marketplace" (Press release). Wildlife Conservation Society. February 9, 2004. Retrieved October 13, 2016.

151. "Greatest threat to Asia's wildlife is hunting, scientists say" (Press release). Wildlife Conservation Society. April 9, 2002. Retrieved October 13, 2016.

152. Hance, Jeremy (January 19, 2009). "Wildlife trade creating 'empty forest syndrome' across the globe". Mongabay.

153. Rodrigues, Ana S. L.; Andelman, Sandy J.; Bakarr, Mohamed I.; Boitani, Luigi; Brooks, Thomas M.; Cowling, Richard M.; Fishpool, Lincoln D. C.; da Fonseca, Gustavo A. B.; Gaston, Kevin J.; Hoffmann, Michael; Long, Janice S.; Marquet, Pablo A.; Pilgrim, John D.; Pressey, Robert L.; Schipper, Jan; Sechrest, Wes; Stuart, Simon N.; Underhill, Les G.; Waller, Robert W.; Watts, Matthew E. J.; Yan, Xie (2004). "Effectiveness of the global protected area network in representing species diversity". Nature. 428 (6983): 640–3. Bibcode:2004Natur.428..640R. doi:10.1038/nature02422. PMID 15071592.

154. Wilcove, David S; Wikelski, Martin (2008). "Going, Going, Gone: Is Animal Migration Disappearing". PLoS Biology. 6 (7): e188. doi:10.1371/journal.pbio.0060188. PMC 2486312 . PMID 18666834.

155. Becker, C. G.; Fonseca, C. R.; Haddad, C. F. B.; Batista, R. F.; Prado, P. I. (2007). "Habitat Split and the Global Decline of Amphibians". Science. 318 (5857): 1775–7. Bibcode:2007Sci...318.1775B. doi:10.1126/science.1149374. PMID 18079402.

156. Schmidt, Gerald (2005). "Ecology & Anthropology: A Field Without Future?". Ecological and Environmental Anthropology. 1 (1): 13–5. OCLC 729066337.

157. McMenamin, S. K.; Hadly, E. A.; Wright, C. K. (2008). "Climatic change and wetland desiccation cause amphibian decline in Yellowstone National Park". Proceedings of the National Academy of Sciences. 105 (44): 16988–93. Bibcode:2008PNAS..10516988M. doi:10.1073/pnas.0809090105. PMC 2579365 . PMID 18955700.

158. Wyman, Richard L. (1991). Global climate change and life on earth. New York: Routledge, Chapman and Hall. ISBN 0-412-02821-2.

159. Thomas, Chris D.; Cameron, Alison; Green, Rhys E.; Bakkenes, Michel; Beaumont, Linda J.; Collingham, Yvonne C.; Erasmus, Barend F. N.; de Siqueira, Marinez Ferreira; Grainger, Alan; Hannah, Lee; Hughes, Lesley; Huntley, Brian; van Jaarsveld, Albert S.; Midgley, Guy F.; Miles, Lera; Ortega-Huerta, Miguel A.; Townsend Peterson, A.; Phillips, Oliver L.; Williams, Stephen E. (2004). "Extinction risk from climate change". Nature. 427 (6970): 145–8. Bibcode:2004Natur.427..145T. doi:10.1038/nature02121. PMID 14712274. Lay summary – National Geographic News (July 12, 2004).

Effects of Global Warming

Chapter Contents

The global warming effect describes what could happen if the world's climate warms. In effect, the global warming effect states that excess greenhouse gasses released into the air will trap the sun's heat in the earth's atmosphere. The accumulation of that heat will warm the earth, causing climates to change. Climate change can cause devastating effects on the planet and those who live in it. Such effects can take the form of rising ocean levels and stronger storms.

INTRODUCTION

The effects of global warming are the environmental and social changes caused (directly or indirectly) by human emissions of greenhouse gases. There is a scientific consensus that climate change is occurring, and that human activities are the primary driver. Many impacts of climate change have already been observed, including glacier retreat, changes in the timing of seasonal events (e.g., earlier flowering of plants), and changes in agricultural productivity.

Future effects of climate change will vary depending on climate change policies and social development. The two main policies to address climate change are reducing human greenhouse gas emissions (climate change mitigation) and adapting to the impacts of climate change. Geoengineering is another policy option.

Near-term climate change policies could significantly affect long-term climate change impacts. Stringent mitigation policies might be able to limit global warming (in 2100) to around 2 °C or below, relative to pre-industrial levels. Without mitigation, increased energy demand and extensive use of fossil fuels might lead to global warming of around 4 °C. Higher magnitudes of global warming would be more difficult to adapt to, and would increase the risk of negative impacts.

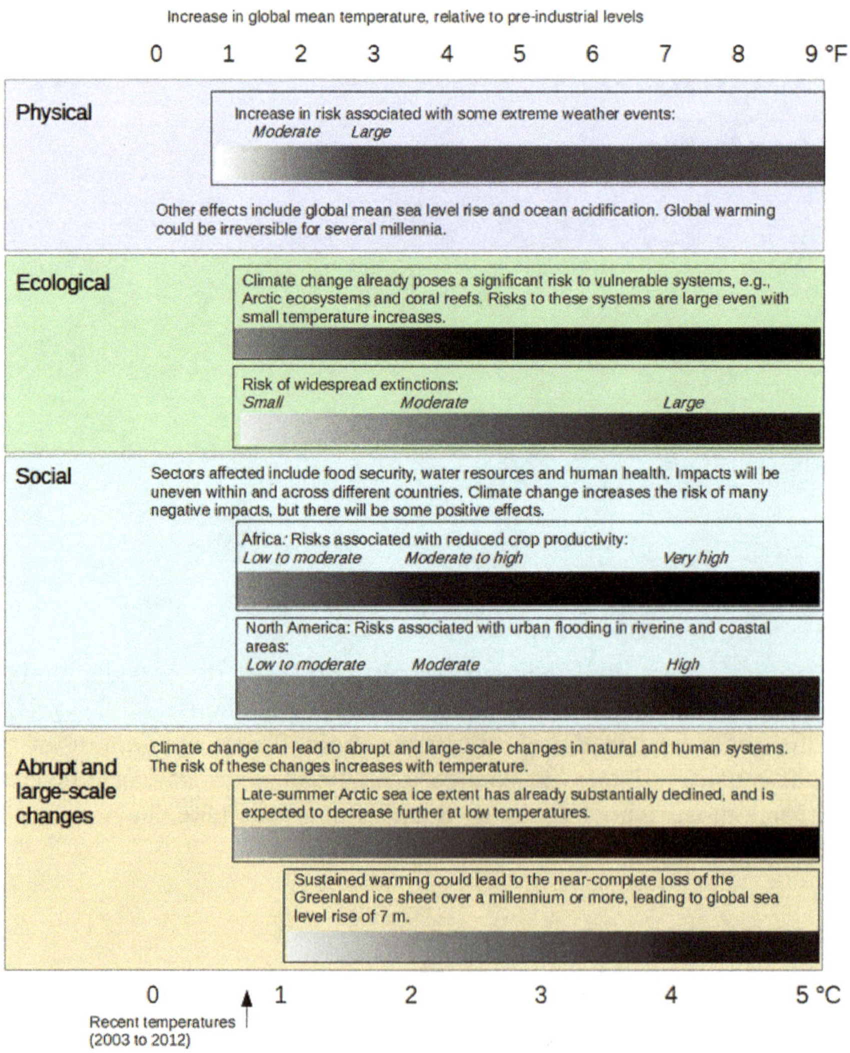

Summary of climate change impacts.

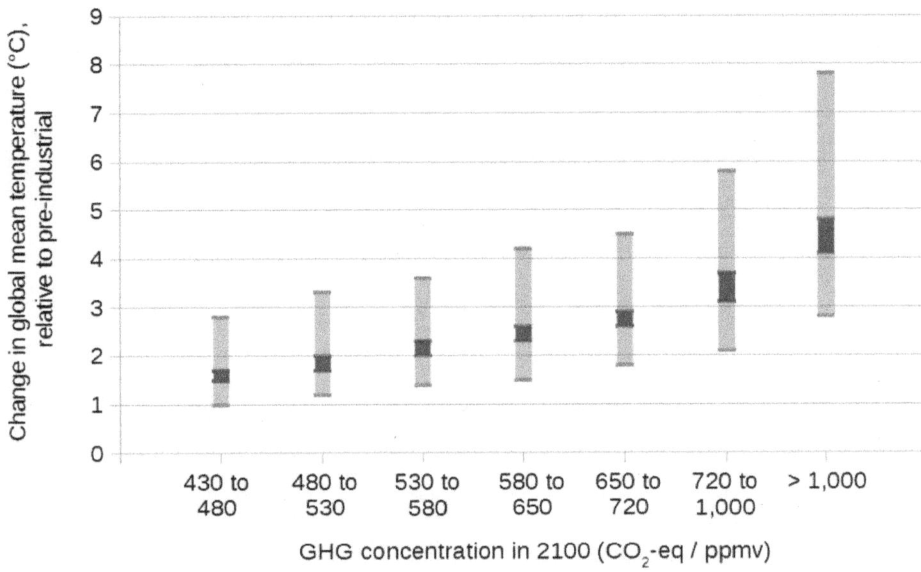

Projected global warming in 2100 for a range of emission scenarios.

Definitions

In this article, "climate change" means a change in climate that persists over a sustained period of time. The World Meteorological Organization defines this time period as 30 years. Examples of climate change include increases in global surface temperature (global warming), changes in rainfall patterns, and changes in the frequency of extreme weather events. Changes in climate may be due to natural causes, e.g., changes in the sun's output, or due to human activities, e.g., changing the composition of the atmosphere. Any human-induced changes in climate will occur against a background of natural climatic variations and of variations in human activity such as population growth on shores or in arid areas which increase or decrease climate vulnerability.

Also, the term "anthropogenic forcing" refers to the influence exerted on a habitat or chemical environment by humans, as opposed to a natural process.

Temperature Changes

The graph above shows the average of a set of temperature simulations for the 20th century (black line), followed by projected temperatures for the 21st century based on three greenhouse gas emission scenarios (colored lines).

This article breaks down some of the impacts of climate change according to different levels of future global warming. This way of describing impacts has, for instance, been used in the IPCC (Intergovernmental Panel on Climate Change) Assessment Reports on climate change. The instrumental temperature record shows global warming of around 0.6 °C during the 20th century.

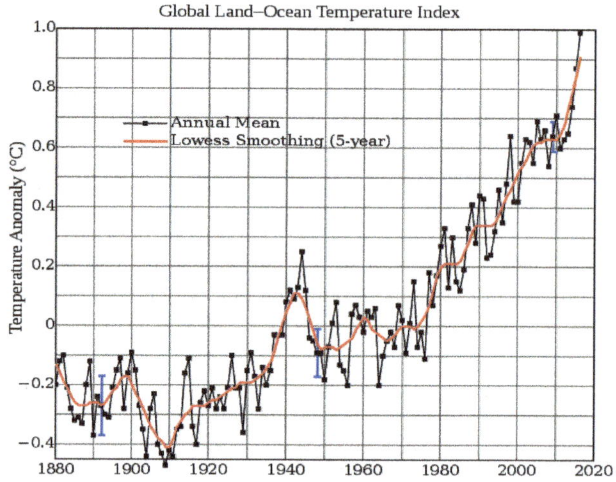

Global mean surface temperature change since 1880, relative to the 1951–1980 mean. Source: NASA GISS

SRES Emissions Scenarios

The future level of global warming is uncertain, but a wide range of estimates (projections) have been made. The IPCC's "SRES" scenarios have been frequently used to make projections of future climate change. The SRES scenarios are "baseline" (or "reference") scenarios, which means that they do not take into account any current or future measures to limit GHG emissions (e.g., the UNFCCC's Kyoto Protocol and the Cancún agreements). Emissions projections of the SRES scenarios are broadly comparable in range to the baseline emissions scenarios that have been developed by the scientific community.

In the IPCC Fourth Assessment Report, changes in future global mean temperature were projected using the six SRES "marker" emissions scenarios. Emissions projections for the six SRES "marker" scenarios are representative of the full set of forty SRES scenarios. For the lowest emissions SRES marker scenario ("B1" - see the SRES article for details on this scenario), the best estimate for global mean temperature is an increase of 1.8 °C (3.2 °F)by the end of the 21st century. This projection is relative to global temperatures at the end of the 20th century. The "likely" range (greater than 66% probability, based on expert judgement) for the SRES B1 marker scenario is 1.1–2.9 °C (2.0–5.2 °F). For the highest emissions SRES marker scenario (A1FI), the best estimate for global mean temperature increase is 4.0 °C (7.2 °F), with a "likely" range of 2.4–6.4 °C (4.3–11.5 °F).

The range in temperature projections partly reflects (1) the choice of emissions scenario, and (2) the "climate sensitivity". For (1), different scenarios make different assumptions of future social and economic development (e.g., economic growth, population level, energy policies), which in turn affects projections of greenhouse gas (GHG) emissions. The projected magnitude of warming by 2100 is closely related to the level of cumulative emissions over the 21st century (i.e. total emissions between 2000-2100). The higher the cumulative emissions over this time period, the greater the level of warming is projected to occur.

(2) reflects uncertainty in the response of the climate system to past and future GHG emissions, which is measured by the climate sensitivity). Higher estimates of climate sensitivity lead to greater projected warming, while lower estimates of climate sensitivity lead to less projected warming.

Over the next several millennia, projections suggest that global warming could be irreversible. Even if emissions were drastically reduced, global temperatures would remain close to their highest level for at least 1,000 years (see the later section on irreversibilities).

Projected Warming in Context

Scientists have used various "proxy" data to assess past changes in Earth's climate (paleoclimate). Sources of proxy data include historical records (such as farmers' logs), tree rings, corals, fossil pollen, ice cores, and ocean and lake sediments. Analysis of these data suggest that recent warming is unusual in the past 400 years, possibly longer. By the end of the 21st century, temperatures may increase to a level not experienced since the mid-Pliocene, around 3 million years ago. At that time, models suggest that mean global temperatures were about 2–3 °C warmer than pre-industrial temperatures. Even a 2 °C rise above the pre-industrial level would be outside the range of temperatures experienced by human civilization.

Physical Impacts

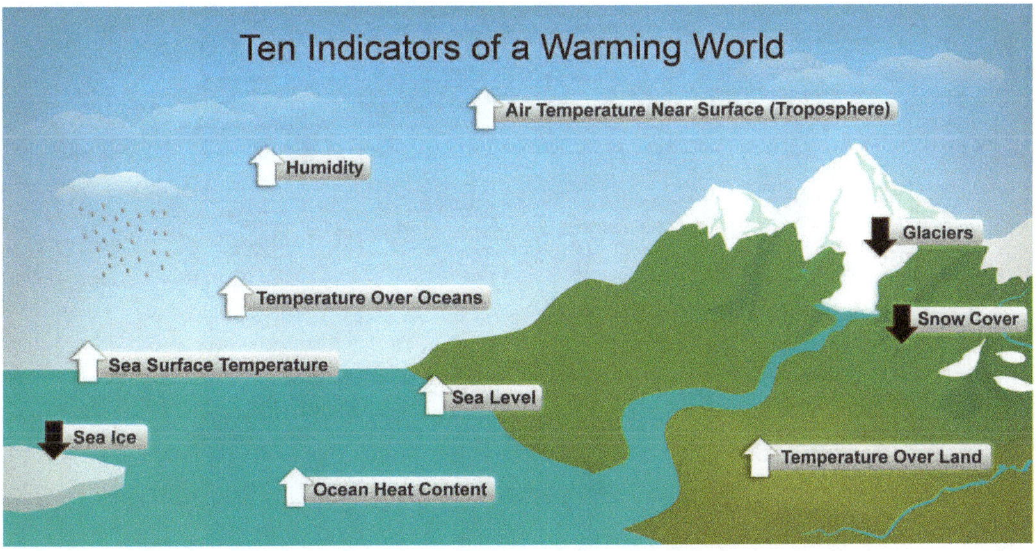

Seven of these indicators would be expected to increase in a warming world and observations show that they are, in fact, increasing. Three would be expected to decrease and they are, in fact, decreasing.

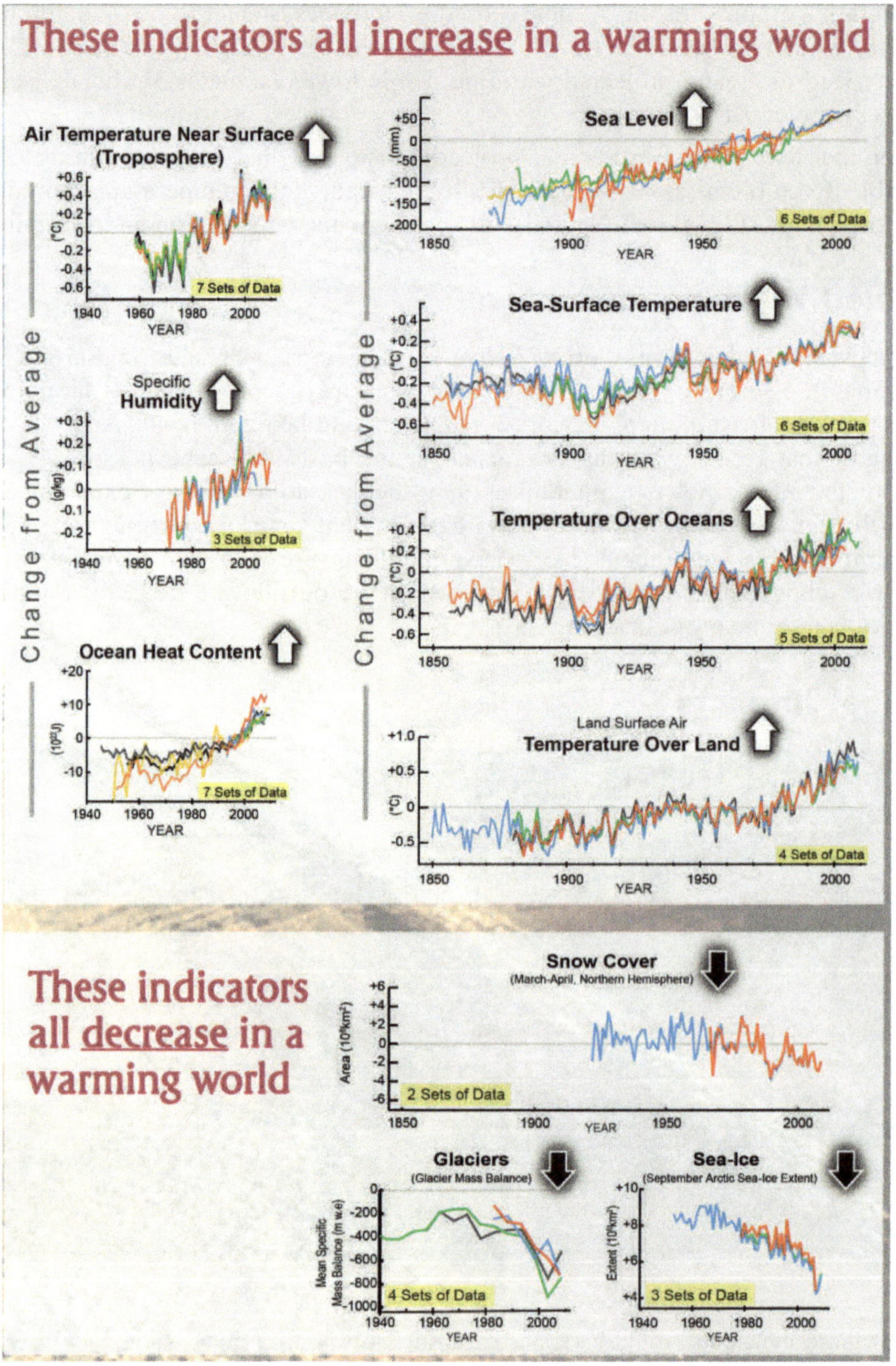

This set of graphs show changes in climate indicators over several decades. Each of the different colored lines in each panel represents an independently analyzed set of data. The data come from many different technologies including weather stations, satellites, weather balloons, ships and buoys.

A broad range of evidence shows that the climate system has warmed. Evidence of global warming is shown in the graphs opposite. Some of the graphs show a positive trend, e.g., increasing temperature over land and the ocean, and sea level rise. Other graphs show a negative trend, e.g., decreased snow cover in the Northern Hemisphere, and declining Arctic sea ice extent. Evidence of warming is also apparent in living (biological) systems.

Human activities have contributed to a number of the observed changes in climate. This contribution has principally been through the burning of fossil fuels, which has led to an increase in the concentration of GHGs in the atmosphere. Another human influence on the climate are sulfur dioxide emissions, which are a precursor to the formation of sulfate aerosols in the atmosphere.

Human-induced warming could lead to large-scale, irreversible, and/or abrupt changes in physical systems. An example of this is the melting of ice sheets, which contributes to sea level rise. The probability of warming having unforeseen consequences increases with the rate, magnitude, and duration of climate change.

Effects on Weather

Observations show that there have been changes in weather. As climate changes, the probabilities of certain types of weather events are affected.

Projected change in annual average precipitation by the end of the 21st century, based on a medium emissions scenario (SRES A1B) (Credit: NOAA Geophysical Fluid Dynamics Laboratory).

Changes have been observed in the amount, intensity, frequency, and type of precipitation. Widespread increases in heavy precipitation have occurred, even in places where total rain amounts have decreased. With medium confidence (see footnote 1), IPCC (2012) concluded that human influences had contributed to an increase in heavy precipitation events at the global scale.

Projections of future changes in precipitation show overall increases in the global average, but with substantial shifts in where and how precipitation falls.[:24] Projections suggest a reduction in rainfall in the subtropics, and an increase in precipitation in subpolar latitudes and some equatorial regions. In other words, regions which are dry at present will in general become even drier, while regions that are currently wet will in general become even wetter. This projection does not apply to every locale, and in some cases can be modified by local conditions.

Extreme Weather

Over most land areas since the 1950s, it is very likely that there have been fewer or warmer cold days and nights. Hot days and nights have also very likely become warmer or more frequent. Human activities have very likely contributed to these trends. There may have been changes in other climate extremes (e.g., floods, droughts and tropical cyclones) but these changes are more difficult to identify.

Projections suggest changes in the frequency and intensity of some extreme weather events. Confidence in projections varies over time.

Near-Term Projections (2016–2035)

Some changes (e.g., more frequent hot days) will probably be evident in the near term, while other near-term changes (e.g., more intense droughts and tropical cyclones) are more uncertain.

Long-Term Projections (2081–2100)

Future climate change will be associated with more very hot days and fewer very cold days. The frequency, length and intensity of heat waves will very likely increase over most land areas. Higher growth in anthropogenic GHG emissions will be associated with larger increases in the frequency and severity of temperature extremes.

Assuming high growth in GHG emissions (IPCC scenario RCP8.5), presently dry regions may be affected by an increase in the risk of drought and reductions in soil moisture. Over most of the mid-latitude land masses and wet tropical regions, extreme precipitation events will very likely become more intense and frequent.

Tropical Cyclones

At the global scale, the frequency of tropical cyclones will probably decrease or be unchanged. Global mean tropical cyclone maximum wind speed and precipitation rates will likely increase. Changes in tropical cyclones will probably vary by region, but these variations are uncertain.

Effects of Climate Extremes

The impacts of extreme events on the environment and human society will vary. Some impacts will be beneficial—e.g., fewer cold extremes will probably lead to fewer cold deaths. Overall, however, impacts will probably be mostly negative.

Cryosphere

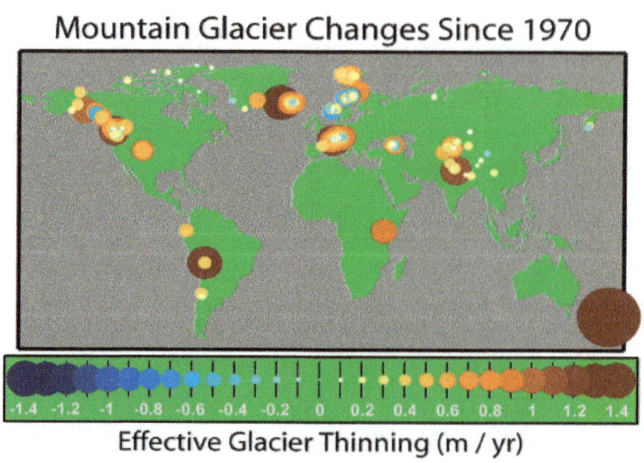

A map of the change in thickness of mountain glaciers since 1970. Thinning in orange and red, thickening in blue.

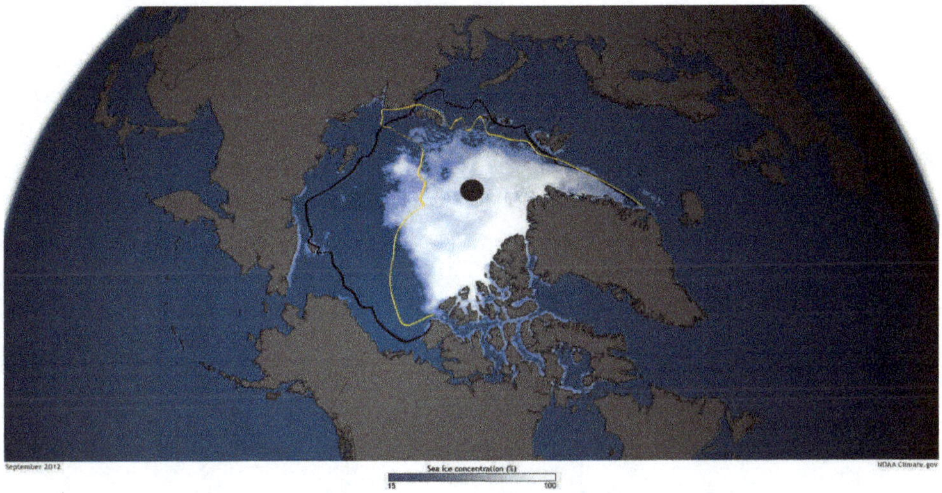

A map that shows ice concentration on 16 September 2012, along with the extent of the previous record low (yellow line) and the mid-September median extent (black line) setting a new record low that was 18 percent smaller than the previous record and nearly 50 percent smaller than the long-term (1979-2000) average.

The cryosphere is made up of areas of the Earth which are covered by snow or ice. Observed changes in the cryosphere include declines in Arctic sea ice extent, the widespread retreat of alpine glaciers,and reduced snow cover in the Northern Hemisphere.

Solomon *et al.* (2007) assessed the potential impacts of climate change on summertime Arctic sea ice extent. Assuming high growth in greenhouse gas emissions (SRES A2), some models projected that Arctic sea ice in the summer could largely disappear by the end of the 21st century. More recent projections suggest that the Arctic summers could be ice-free (defined as ice extent less than 1 million square km) as early as 2025-2030.

During the 21st century, glaciers and snow cover are projected to continue their widespread retreat.In the western mountains of North America, increasing temperatures and changes in precipitation are projected to lead to reduced snowpack. Snowpack is the seasonal accumulation of slow-melting snow. The melting of the Greenland and West Antarctic ice sheets could contribute to sea level rise, especially over long time-scales (see the section on Greenland and West Antarctic Ice sheets).

Changes in the cryosphere are projected to have social impacts. For example, in some regions, glacier retreat could increase the risk of reductions in seasonal water availability. Barnett *et al.* (2005)estimated that more than one-sixth of the world's population rely on glaciers and snowpack for their water supply.

Oceans

The role of the oceans in global warming is complex. The oceans serve as a sink for carbon dioxide, taking up much that would otherwise remain in the atmosphere, but increased levels of CO_2 have led to ocean acidification. Furthermore, as the temperature of the oceans increases, they become less able to absorb excess CO

2. The ocean have also acted as a sink in absorbing extra heat from the atmosphere.[4] The increase in ocean heat content is much larger than any other store of energy in the Earth's heat balance over the two periods 1961 to 2003 and 1993 to 2003, and accounts for more than 90% of the possible increase in heat content of the Earth system during these periods.

Global warming is projected to have a number of effects on the oceans. Ongoing effects include rising sea levels due to thermal expansion and melting of glaciers and ice sheets, and warming of the ocean surface, leading to increased temperature stratification. Other possible effects include large-scale changes in ocean circulation.

Acidification

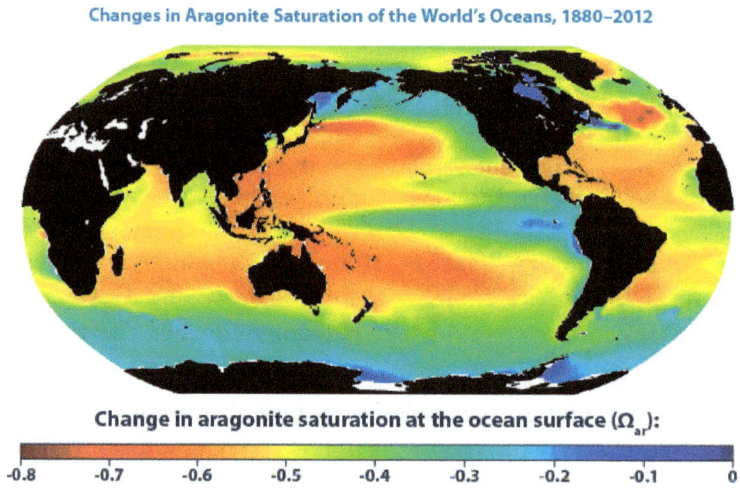

This map shows changes in the amount of aragonite dissolved in ocean surface waters between the 1880s and the most recent decade (2003-2012). Historical modeling suggests that since the 1880s, increased CO_2 has led to lower aragonite saturation levels (less availability of minerals) in the oceans around the world. The largest decreases in aragonite saturation have occurred in tropical waters. However, decreases in cold areas may be of greater concern because colder waters typically have lower aragonite levels to begin with.

About one-third of the carbon dioxide emitted by human activity has already been taken up by the oceans. As carbon dioxide dissolves in sea water, carbonic acid is formed, which has the effect of acidifying the ocean, measured as a change in pH. The uptake of human carbon emissions since the year 1750 has led to an average decrease in pH of 0.1 units. Projections using the SRES emissions scenarios suggest a further reduction in average global surface ocean pH of between 0.14 and 0.35 units over the 21st century.

The effects of ocean acidification on the marine biosphere have yet to be documented. Laboratory experiments suggest beneficial effects for a few species, with potentially highly detrimental effects for a substantial number of species. With medium confidence, Fischlin *et al.* (2007) projected that future ocean acidification and climate change would impair a wide range of planktonic and shallow benthic marine organisms that use aragonite to make their shells or skeletons, such as corals and marine snails (pteropods), with significant impacts particularly in the Southern Ocean.

Oxygen Depletion

The amount of oxygen dissolved in the oceans may decline, with adverse consequences for ocean life.

Sea Level Rise

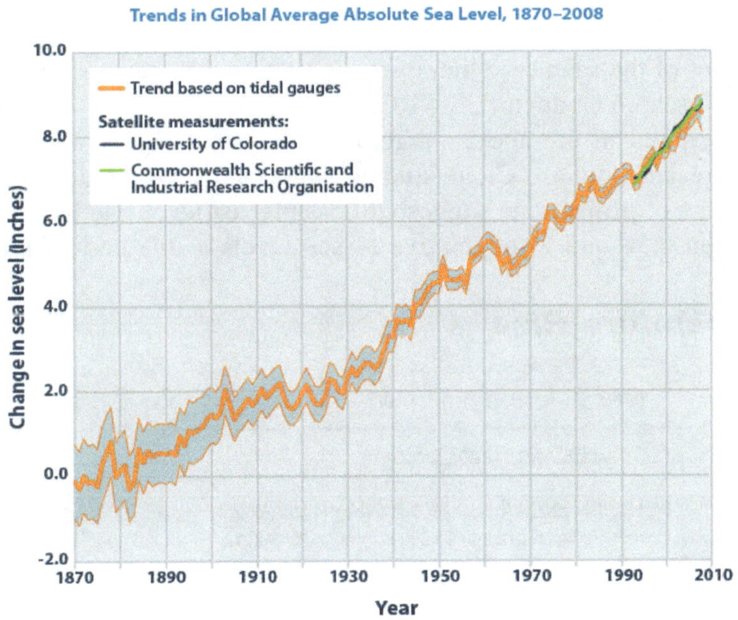

Trends in global average absolute sea level, 1870-2008.

There is strong evidence that global sea level rose gradually over the 20th century. With high confidence, Bindoff *et al.* (2007) concluded that between the mid-19th and mid-20th centuries, the rate of sea level rise increased. Authors of the IPCC Fourth Assessment Synthesis Report (IPCC AR4 SYR, 2007) reported that between the years 1961 and 2003, global average sea level rose at an average rate of 1.8 mm per year (mm/yr), with a range of 1.3–2.3 mm/yr. Between 1993 and 2003, the rate increased above the previous period to 3.1 mm/yr (range of 2.4–3.8 mm/yr). Authors of IPCC AR4 SYR (2007) were uncertain whether the increase in rate from 1993 to 2003 was due to natural variations in sea level over the time period, or whether it reflected an increase in the underlying long-term trend.

There are two main factors that have contributed to observed sea level rise. The first is thermal expansion: as ocean water warms, it expands. The second is from the contribution of land-based ice due to increased melting. The major store of water on land is found in glaciers and ice sheets. Anthropogenic forces very likely (greater than 90% probability, based on expert judgement) contributed to sea level rise during the latter half of the 20th century.

There is a widespread consensus that substantial long-term sea level rise will continue for centuries to come. In their Fourth Assessment Report, the IPCC projected sea level rise to the end of the 21st century using the SRES emissions scenarios. Across the six SRES marker scenarios, sea level was projected to rise by 18 to 59 cm (7.1 to 23.2 in), relative to sea level at the end of the 20th century. Thermal expansion is the largest component in these projections, contributing 70-75% of the central estimate for all scenarios. Due to a lack of scientific understanding, this sea level rise estimate does not include all of the possible contributions of ice sheets (see the section on Greenland and West Antarctic Ice sheets).

An assessment of the scientific literature on climate change was published in 2010 by the US National Research Council (US NRC, 2010). NRC (2010) described the projections in AR4 (i.e. those cited in the above paragraph) as "conservative", and summarized the results of more recent studies. Cited studies suggested a great deal of uncertainty in projections. A range of projections suggested possible sea level rise by the end of the 21st century of between 0.56 and 2 m, relative to sea levels at the end of the 20th century.

Ocean Temperature Rise

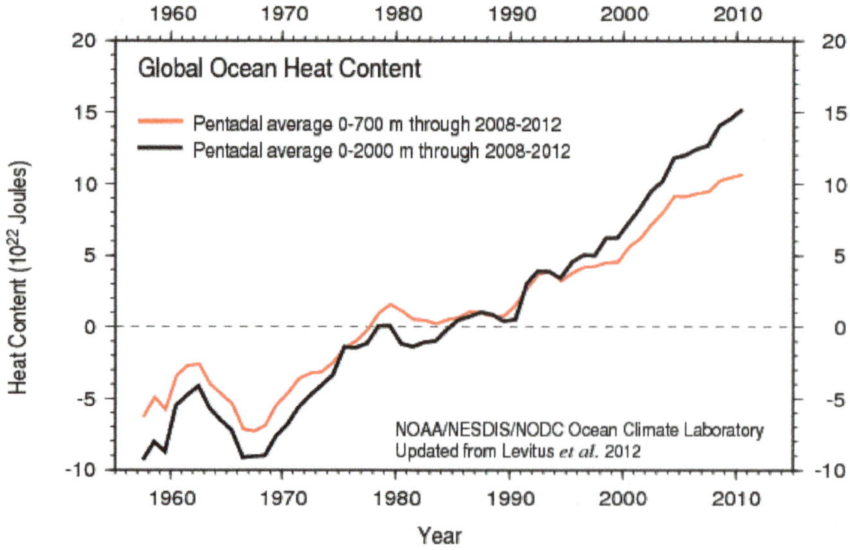

Global ocean heat content from 1955-2012

From 1961 to 2003, the global ocean temperature has risen by 0.10 °C from the surface to a depth of 700 m. There is variability both year-to-year and over longer time scales, with global ocean heat content observations showing high rates of warming for 1991–2003, but some cooling from 2003 to 2007. The temperature of the Antarctic

Southern Ocean rose by 0.17 °C (0.31 °F) between the 1950s and the 1980s, nearly twice the rate for the world's oceans as a whole. As well as having effects on ecosystems (e.g. by melting sea ice, affecting algae that grow on its underside), warming reduces the ocean's ability to absorb CO_2. It is likely (greater than 66% probability, based on expert judgement) that anthropogenic forcing contributed to the general warming observed in the upper several hundred metres of the ocean during the latter half of the 20th century.

Regions

Temperatures across the world in the 1880s (left) and the 1980s (right), as compared to average temperatures from 1951 to 1980.

Projected changes in average temperatures across the world in the 2050s under three greenhouse gas (GHG) emissions scenarios.

Regional effects of global warming vary in nature. Some are the result of a generalised global change, such as rising temperature, resulting in local effects, such as melting ice. In other cases, a change may be related to a change in a particular ocean current or weather system. In such cases, the regional effect may be disproportionate and will not necessarily follow the global trend.

There are three major ways in which global warming will make changes to regional climate: melting or forming ice, changing the hydrological cycle (of evaporation and precipitation) and changing currents in the oceans and air flows in the atmosphere. The coast can also be considered a region, and will suffer severe impacts from sea level rise.

Observed Impacts

With very high confidence, Rosenzweig *et al.* (2007) concluded that physical and biological systems on all continents and in most oceans had been affected by recent climate changes, particularly regional temperature increases. Impacts include earlier leafing of trees and plants over many regions; movements of species to higher latitudes and altitudes in the Northern Hemisphere; changes in bird migrations in Europe, North America, and Australia; and shifting of the oceans' plankton and fish from cold- to warm-adapted communities.

The human influence on the climate can be seen in the geographical pattern of observed warming, with greater temperature increases over land and in polar regions rather than over the oceans. Using models, it is possible to identify the human "signal" of global warming over both land and ocean areas.

Projected Impacts

Projections of future climate changes at the regional scale do not hold as high a level of scientific confidence as projections made at the global scale. It is, however, expected that future warming will follow a similar geographical pattern to that seen already, with greatest warming over land and high northern latitudes, and least over the Southern Ocean and parts of the North Atlantic Ocean. Nearly all land areas will very likely warm more than the global average.

The Arctic, Africa, small islands and Asian megadeltas are regions that are likely to be especially affected by climate change. Low-latitude, less-developed areas are at most risk of experiencing negative impacts due to climate change. Developed countries are also vulnerable to climate change. For example, developed countries will be negatively affected by increases in the severity and frequency of some extreme weather events, such as heat waves. In all regions, some people can be particularly at risk from climate change, such as the poor, young children and the elderly.

Social Systems

The impacts of climate change can be thought of in terms of sensitivity and vulnerability. "Sensitivity" is the degree to which a particular system or sector might be affected, positively or negatively, by climate change and/or climate variability. "Vulnerability" is the degree to which a particular system or sector might be adversely affected by climate change.

The sensitivity of human society to climate change varies. Sectors sensitive to climate change include water resources, coastal zones, human settlements, and human health. Industries sensitive to climate change include agriculture, fisheries, forestry, energy, construction, insurance, financial services, tourism, and recreation.

Food Supply

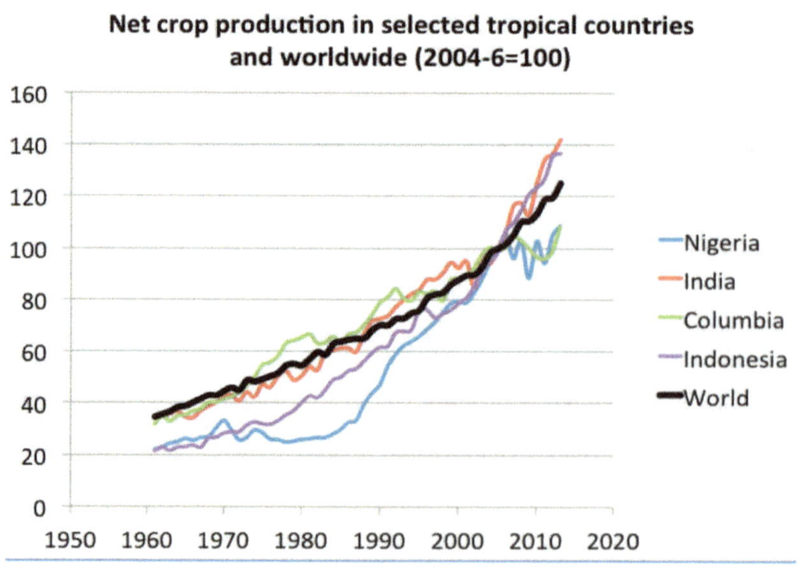

Graph of net crop production worldwide and in selected tropical countries. Raw data from the United Nations.

Climate change will impact agriculture and food production around the world due to: the effects of elevated CO_2 in the atmosphere, higher temperatures, altered precipitation and transpiration regimes, increased frequency of extreme events, and modified weed, pest, and pathogen pressure. In general, low-latitude areas are at most risk of having decreased crop yields.

As of 2007, the effects of regional climate change on agriculture have been small. Changes in crop phenology provide important evidence of the response to recent regional climate change. Phenology is the study of natural phenomena that recur periodically, and how these phenomena relate to climate and seasonal changes. A significant advance in phenology has been observed for agriculture and forestry in large parts of the Northern Hemisphere.

Projections

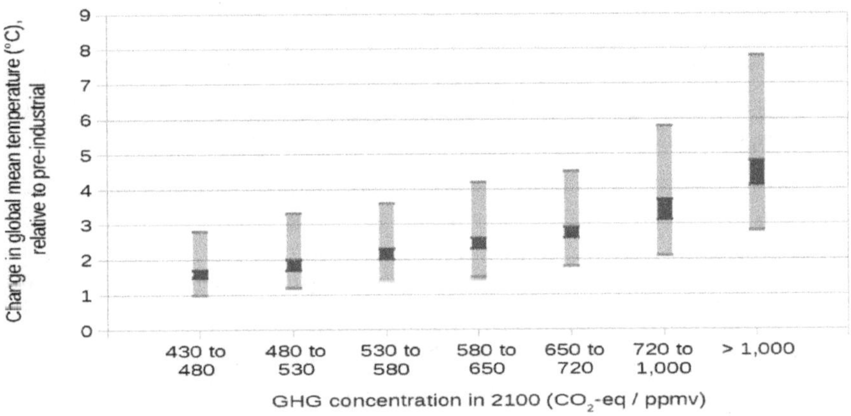

Projected changes in crop yields at different latitudes with global warming. This graph is based on several studies.

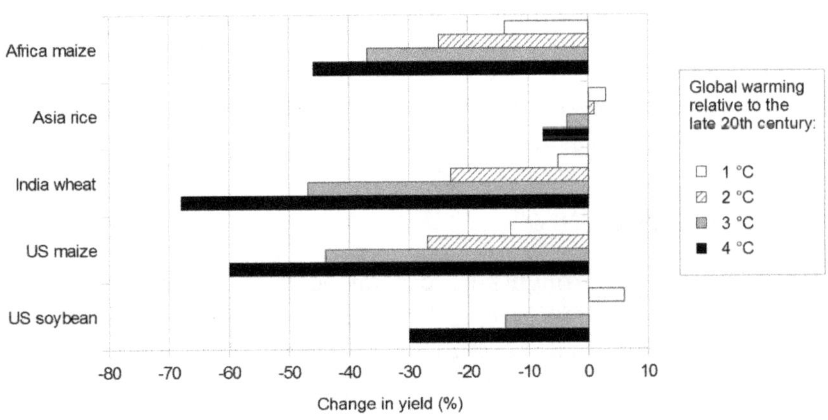

Projected changes in yields of selected crops with global warming. This graph is based on several studies.

With low to medium confidence, Schneider *et al.* (2007) projected that for about a 1 to 3 °C increase in global mean temperature (by the years 2090-2100, relative to average temperatures in the years 1990–2000), there would be productivity decreases for some cereals in low latitudes, and productivity increases in high latitudes. With medium confidence, global production potential was projected to:

- increase up to around 3 °C,
- very likely decrease above about 3 °C.

Most of the studies on global agriculture assessed by Schneider *et al.* (2007) had not incorporated a number of critical factors, including changes in extreme events, or the spread of pests and diseases. Studies had also not considered the development of specific practices or technologies to aid adaptation to climate change.

The graphs opposite show the projected effects of climate change on selected crop yields. Actual changes in yields may be above or below these central estimates.

The projections above can be expressed relative to pre-industrial (1750) temperatures. 0.6 °C of warming is estimated to have occurred between 1750 and 1990-2000. Add 0.6 °C to the above projections to convert them from a 1990-2000 to pre-industrial baseline.

Food Security

Easterling *et al.* (2007) assessed studies that made quantitative projections of climate change impacts on food security. It was noted that these projections were highly uncertain and had limitations. However, the assessed studies suggested a number of fairly robust findings. The first was that climate change would likely increase the number of people at risk of hunger compared with reference scenarios with no climate change. Climate change impacts depended strongly on projected future social and economic development. Additionally, thc magnitudc of climate change impacts was projected to be smaller compared to the impact of social and economic development. In 2006, the global estimate for the number of people undernourished was 820 million. Under the SRES A1, B1, and B2 scenarios (see the SRES article for information on each scenario group), projections for the year 2080 showed a reduction in the number of people undernourished of about 560-700 million people, with a global total of undernourished people of 100-240 million in 2080. By contrast, the SRES A2 scenario showed only a small decrease in the risk of hunger from 2006 levels. The smaller reduction under A2 was attributed to the higher projected future population level in this scenario.

Droughts and Agriculture

Some evidence suggests that droughts have been occurring more frequently because of global warming and they are expected to become more frequent and intense in Africa, southern Europe, the Middle East, most of the Americas, Australia, and Southeast Asia. However, other research suggests that there has been little change in drought over the past 60 years. Their impacts are aggravated because of increased water demand, population growth, urban expansion, and environmental protection efforts in many areas. Droughts result in crop failures and the loss of pasture grazing land for livestock.

Health

Human beings are exposed to climate change through changing weather patterns (temperature, precipitation, sea-level rise and more frequent extreme events) and indirectly through changes in water, air and food quality and changes in ecosystems, agriculture, industry and settlements and the economy (Confalonieri *et al.*, 2007:393). According to an assessment of the scientific literature by Confalonieri *et al.* (2007:393), the effects of climate change to date have been small, but are projected to progressively increase in all countries and regions.

A study by the World Health Organization (WHO, 2009) estimated the effect of climate change on human health. Not all of the effects of climate change were included in their estimates, for example, the effects of more frequent and extreme storms were excluded. Climate change was estimated to have been responsible for 3% of diarrhoea, 3% of malaria, and 3.8% of dengue fever deaths worldwide in 2004. Total attributable mortality was about 0.2% of deaths in 2004; of these, 85% were child deaths.

Projections

With high confidence, authors of the IPCC AR4 Synthesis report projected that climate change would bring some benefits in temperate areas, such as fewer deaths from cold exposure, and some mixed effects such as changes in range and transmission potential of malaria in Africa. Benefits were projected to be outweighed by negative health effects of rising temperatures, especially in developing countries.

With very high confidence, Confalonieri *et al.* (2007) concluded that economic development was an important component of possible adaptation to climate change. Economic growth on its own, however, was not judged to be sufficient to insulate the world's population from disease and injury due to climate change. Future vulnerability to climate change will depend not only on the extent of social and economic change, but also on how the benefits and costs of change are distributed in society. For example, in the 19th century, rapid urbanization in western Europe lead to a plummeting in population health. Other factors important in determining the health of populations include education, the availability of health services, and public-health infrastructure.

Water Resources

A number of climate-related trends have been observed that affect water resources. These include changes in precipitation, the crysosphere and surface waters (e.g., changes in river flows). Observed and projected impacts of climate change on freshwater systems and their management are mainly due to changes in temperature, sea level and precipitation variability. Sea level rise will extend areas of salinization of groundwater and estuaries, resulting in a decrease in freshwater availability for humans and ecosystems in coastal areas. In an assessment of the scientific literature, Kundzewicz *et al.* (2007) concluded, with high confidence, that:

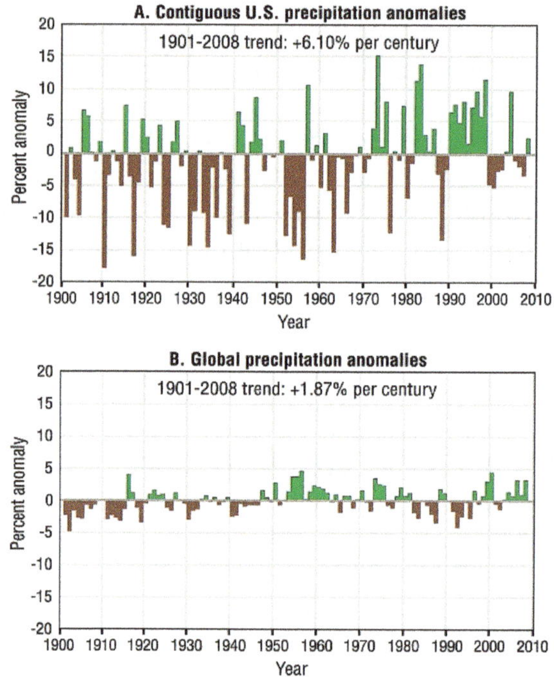

Precipitation during the 20th century and up through 2008 during global warming, the NOAA estimating an observed trend over that period of 1.87% global precipitation increase per century.

- The negative impacts of climate change on freshwater systems outweigh the benefits. All of the regions assessed in the IPCC Fourth Assessment Report (Africa, Asia, Australia and New Zealand, Europe, Latin America, North America, Polar regions (Arctic and Antarctic), and small islands) showed an overall net negative impact of climate change on water resources and freshwater ecosystems.
- Semi-arid and arid areas are particularly exposed to the impacts of climate change on freshwater. With very high confidence, it was judged that many of these areas, e.g., the Mediterranean basin, Western United States, Southern Africa, and north-eastern Brazil, would suffer a decrease in water resources due to climate change.

Migration and Conflict

General circulation models project that the future climate change will bring wetter coasts, drier mid-continent areas, and further sea level rise. Such changes could result in the gravest effects of climate change through human migration. Millions might be displaced by shoreline erosions, river and coastal flooding, or severe drought.

Migration related to climate change is likely to be predominantly from rural areas in developing countries to towns and cities. In the short term climate stress is likely to add incrementally to existing migration patterns rather than generating entirely new flows of people.

It has been argued that environmental degradation, loss of access to resources (e.g., water resources), and resulting human migration could become a source of political and even military conflict. Factors other than climate change may, however, be more important in affecting conflict. For example, Wilbanks *et al.* (2007) suggested that major environmentally influenced conflicts in Africa were more to do with the relative abundance of resources, e.g., oil and diamonds, than with resource scarcity. Scott *et al.* (2001) placed only low confidence in predictions of increased conflict due to climate change.

A 2013 study found that significant climatic changes were associated with a higher risk of conflict worldwide, and predicted that "amplified rates of human conflict could represent a large and critical social impact of anthropogenic climate change in both low- and high-income countries."Similarly, a 2014 study found that higher temperatures were associated with a greater likelihood of violent crime, and predicted that global warming would cause millions of such crimes in the United States alone during the 21st century.

Military planners are concerned that global warming is a "threat multiplier". "Whether it is poverty, food and water scarcity, diseases, economic instability, or threat of natural disasters, the broad range of changing climatic conditions may be far reaching. These challenges may threaten stability in much of the world".

Aggregate Impacts

Aggregating impacts adds up the total impact of climate change across sectors and/or regions. Examples of aggregate measures include economic cost (e.g., changes in gross domestic product (GDP) and the social cost of carbon), changes in ecosystems (e.g., changes over land area from one type of vegetation to another), human health impacts, and the number of people affected by climate change. Aggregate measures such as economic cost require researchers to make value judgements over the importance of impacts occurring in different regions and at different times.

Observed Impacts

Global losses reveal rapidly rising costs due to extreme weather-related events since the 1970s. Socio-economic factors have contributed to the observed trend of global losses, e.g., population growth, increased wealth. Part of the growth is also related to regional climatic factors, e.g., changes in precipitation and flooding events. It is difficult to quantify the relative impact of socio-economic factors and climate change on the observed trend. The trend does, however, suggest increasing vulnerability of social systems to climate change.

Projected Impacts

The total economic impacts from climate change are highly uncertain. With medium confidence, Smith *et al.* (2001) concluded that world GDP would change by plus or minus a few percent for a small increase in global mean temperature (up to around 2 °C relative to the 1990 temperature level). Most studies assessed by Smith *et al.* (2001) projected losses in world GDP for a medium increase in global mean temperature (above 2-3 °C relative to the 1990 temperature level), with increasing losses for greater temperature increases. This assessment is consistent with the findings of more recent studies, as reviewed by Hitz and Smith (2004).

Economic impacts are expected to vary regionally. For a medium increase in global mean temperature (2-3 °C of warming, relative to the average temperature between 1990–2000), market sectors in low-latitude and less-developed areas might experience net costs due to climate change. On the other hand, market sectors in high-latitude and developed regions might experience net benefits for this level of warming. A global mean temperature increase above about 2-3 °C (relative to 1990-2000) would very likely result in market sectors across all regions experiencing either declines in net benefits or rises in net costs.

Aggregate impacts have also been quantified in non-economic terms. For example, climate change over the 21st century is likely to adversely affect hundreds of millions of people through increased coastal flooding, reductions in water supplies, increased malnutrition and increased health impacts.

Biological Systems

Observed Impacts on Biological Systems

A vast array of physical and biological systems across the Earth are being affected by human-induced global warming.

With very high confidence, Rosenzweig *et al.* (2007) concluded that recent warming had strongly affected natural biological systems. Hundreds of studies have documented responses of ecosystems, plants, and animals to the climate changes that have already occurred. For example, in the Northern Hemisphere, species are almost uniformly moving their ranges northward and up in elevation in search of cooler temperatures. Humans are very likely causing changes in regional temperatures to which plants and animals are responding.

Projected Impacts on Biological Systems

By the year 2100, ecosystems will be exposed to atmospheric CO_2 levels substantially higher than in the past 650,000 years, and global temperatures at least among the highest of those experienced in the past 740,000 years. Significant disruptions of ecosystems are projected to increase with future climate change. Examples of disruptions include disturbances such as fire, drought, pest infestation, invasion of species, storms, and coral bleaching events. The stresses caused by climate change, added to other stresses on ecological systems (e.g., land conversion, land degradation, harvesting, and pollution), threaten substantial damage to or complete loss of some unique ecosystems, and extinction of some critically endangered species.

Climate change has been estimated to be a major driver of biodiversity loss in cool conifer forests, savannas, mediterranean-climate systems, tropical forests, in the Arctic tundra, and in coral reefs. In other ecosystems, land-use change may be a stronger driver of biodiversity loss at least in the near-term. Beyond the year 2050, climate change may be the major driver for biodiversity loss globally.

A literature assessment by Fischlin *et al.* (2007) included a quantitative estimate of the number of species at increased risk of extinction due to climate change. With medium confidence, it was projected that approximately 20 to 30% of plant and animal species assessed so far (in an unbiased sample) would likely be at increasingly high risk of extinction should global mean temperatures exceed a warming of 2 to 3 °C above pre-industrial temperature levels. The uncertainties in this estimate, however, are large: for a rise of about 2 °C the percentage may be as low as 10%, or for about 3 °C, as high as 40%, and depending on biota (all living organisms of an area, the flora and fauna considered as a unit) the range is between 1% and 80%. As global average temperature exceeds 4 °C above pre-industrial levels, model projections suggested that there could be significant extinctions (40-70% of species that were assessed) around the globe.

Assessing whether future changes in ecosystems will be beneficial or detrimental is largely based on how ecosystems are valued by human society. For increases in global average temperature exceeding 1.5 to 2.5 °C (relative to global temperatures over the years 1980-1999)and in concomitant atmospheric CO_2 concentrations, projected changes in ecosystems will have predominantly negative consequences for biodiversity and ecosystems goods and services, e.g., water and food supply.

Abrupt or Irreversible Changes

Physical, ecological and social systems may respond in an abrupt, non-linear or irregular way to climate change. This is as opposed to a smooth or regular response. A quantitative entity behaves "irregularly" when its dynamics are discontinuous (i.e., not smooth), nondifferentiable, unbounded, wildly varying, or otherwise ill-defined. Such behaviour is often termed "singular". Irregular behaviour in Earth systems may give rise to certain thresholds, which, when crossed, may lead to a large change in the system.

Some singularities could potentially lead to severe impacts at regional or global scales. Examples of "large-scale" singularities are discussed in the articles on abrupt climate change, climate change feedback and runaway climate change. It is possible that human-induced climate change could trigger large-scale singularities, but the probabilities of triggering such events are, for the most part, poorly understood.

With low to medium confidence, Smith *et al.* (2001) concluded that a rapid warming of more than 3 °C above 1990 levels would exceed thresholds that would lead to large-scale discontinuities in the climate system. Since the assessment by Smith *et al.* (2001), improved scientific understanding provides more guidance for two large-scale singularities: the role of carbon cycle feedbacks in future climate change (discussed below in the section on biogeochemical cycles) and the melting of the Greenland and West Antarctic ice sheets.

Biogeochemical Cycles

Climate change may have an effect on the carbon cycle in an interactive "feedback" process. A feedback exists where an initial process triggers changes in a second process that in turn influences the initial process. A positive feedback intensifies the original process, and a negative feedback reduces it. Models suggest that the interaction of the climate system and the carbon cycle is one where the feedback effect is positive.

Using the A2 SRES emissions scenario, Schneider *et al.* (2007) found that this effect led to additional warming by the years 2090-2100 (relative to the 1990–2000) of 0.1–1.5 °C. This estimate was made with high confidence. The climate projections made in the IPCC Fourth Assessment Report summarized earlier of 1.1–6.4 °C account for this feedback effect. On the other hand, with medium confidence, Schneider *et al.* (2007) commented that additional releases of GHGs were possible from permafrost, peat lands, wetlands, and large stores of marine hydrates at high latitudes.

Greenland and West Antarctic Ice sheets

With medium confidence, authors of AR4 concluded that with a global average temperature increase of 1–4 °C (relative to temperatures over the years 1990–2000), at least a partial deglaciation of the Greenland ice sheet, and possibly the West Antarctic ice sheets would occur. The estimated timescale for partial deglaciation was centuries to millennia, and would contribute 4 to 6 metres (13 to 20 ft) or more to sea level rise over this period.

Atlantic Meridional Overturning Circulation

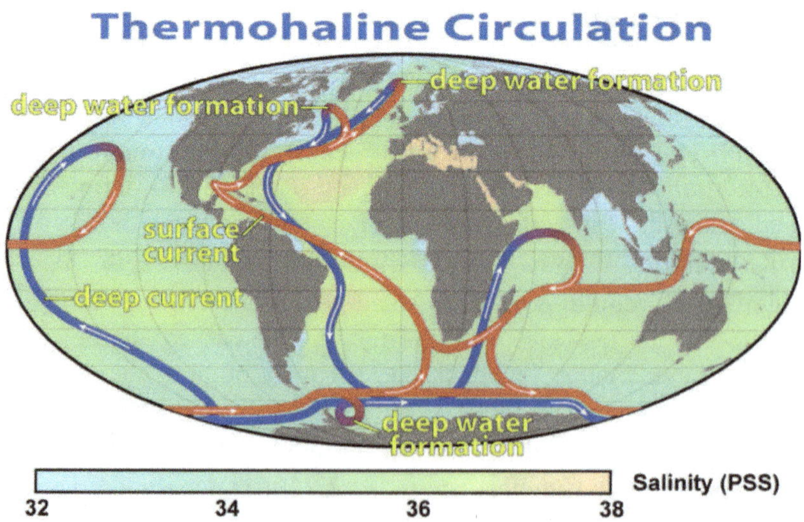

This map shows the general location and direction of the warm surface (red) and cold deep water (blue) currents of the thermohaline circulation. Salinity is represented by color in units of the Practical Salinity Scale. Low values (blue) are less saline, while high values (orange) are more saline.

The Atlantic Meridional Overturning Circulation (AMOC) is an important component of the Earth's climate system, characterized by a northward flow of warm, salty water in the upper layers of the Atlantic and a southward flow of colder water in the deep Atlantic.

The AMOC is equivalently known as the thermohaline circulation (THC). Potential impacts associated with MOC changes include reduced warming or (in the case of abrupt change) absolute cooling of northern high-latitude areas near Greenland and north-western Europe, an increased warming of Southern Hemisphere high-latitudes, tropical drying, as well as changes to marine ecosystems, terrestrial vegetation, oceanic CO_2 uptake, oceanic oxygen concentrations, and shifts in fisheries. According to an assessment by the US Climate Change Science Program (CCSP, 2008b), it is very likely (greater than 90% probability, based on expert judgement) that the strength of the AMOC will decrease over the course of the 21st century. Warming is still expected to occur over most of the European region downstream of the North Atlantic Current in response to increasing GHGs, as well as over North America. Although it is very unlikely (less than 10% probability, based on expert judgement) that the AMOC will collapse in the 21st century, the potential consequences of such a collapse could be severe.

Irreversibilities

Commitment to Radiative Forcing

Emissions of GHGs are a potentially irreversible commitment to sustained radiative forcing in the future. The contribution of a GHG to radiative forcing depends on the gas's ability to trap infrared (heat) radiation, the concentration of the gas in the atmosphere, and the length of time the gas resides in the atmosphere.

CO_2 is the most important anthropogenic GHG. While more than half of the CO_2 emitted is currently removed from the atmosphere within a century, some fraction (about 20%) of emitted CO_2 remains in the atmosphere for many thousands of years. Consequently, CO_2 emitted today is potentially an irreversible commitment to sustained radiative forcing over thousands of years.

This commitment may not be truly irreversible should techniques be developed to remove CO_2 or other GHGs directly from the atmosphere, or to block sunlight to induce cooling. Techniques of this sort are referred to as geoengineering. Little is known about the effectiveness, costs or potential side-effects of geoengineering options. Some geoengineering options, such as blocking sunlight, would not prevent further ocean acidification.

Irreversible impacts

Human-induced climate change may lead to irreversible impacts on physical, biological, and social systems. There are a number of examples of climate change impacts that may be irreversible, at least over the timescale of many human generations. These include the large-scale singularities described above – changes in carbon cycle feedbacks, the melting of the Greenland and West Antarctic ice sheets, and changes to the AMOC. In biological systems, the extinction of species would be an irreversible impact. In social systems, unique cultures may be lost due to climate change. For example, humans living on atoll islands face risks due to sea-level rise, sea-surface warming, and increased frequency and intensity of extreme weather events.

Benefits of global warming

In some locations and industries global warming may increase productivity, though the IPCC cautions that "Estimates agree on the size of the impact (small relative to economic growth), and 17 of the 20 impact estimates shown in Figure 10-1 are negative. Losses accelerate with greater warming, and estimates diverge." The identified benefits are listed below.

CO2 fertilisation effect

CO2 is one of the substances which plants require to grow. Increasing its amount in the air contributes to:

- Improved agriculture in some high latitude regions
- Increased growing season in Greenland
- Increased productivity of sour orange trees
- Increased vegetation activity in high northern latitudes
- Increased plankton biomass in the North Pacific Subtropical Gyre
- Recent increase in forest growth
- Increased Arctic tundra plant reproduction

Human health

- Winter deaths might decline as temperatures warm. However, this is disputed for at least some regions. For example states "Although excess winter deaths evidently do exist, winter cold severity no longer predicts the numbers affected. We conclude that no evidence exists that excess winter deaths in England and Wales will fall if winters warm with climate change."

Ice-free Northwest Passage

- Ships will travel on a shorter route between the Pacific and Atlantic oceans

Animal population changes

Some animals will benefit from the warming:

- Increase in chinstrap and gentoo penguins
- Bigger marmots

Scientific Opinion

The Intergovernmental Panel on Climate Change (IPCC) has published several major assessments on the effects of global warming. Its most recent comprehensive impact assessment was published in 2014. Publications describing the effects of climate change have also been produced by the following organizations:

- American Association for the Advancement of Science (AAAS)
- A report by the Netherlands Environmental Assessment Agency, the Royal Netherlands Meteorological Institute, and Wageningen University and Research Centre
- UK AVOID research programme
- A report by the UK Royal Society and US National Academy of Sciences
- University of New South Wales Climate Change Research Centre
- US National Research Council

A report by Molina *et al.* (no date) states:

The overwhelming evidence of human-caused climate change documents both current impacts with significant costs and extraordinary future risks to society and natural systems

NASA Data and Tools

NASA has released public data and tools to predict how temperature and rainfall patterns worldwide may change through to the year 2100 caused by increasing carbon dioxide in Earth's atmosphere. The dataset shows projected changes worldwide on a regional level simulated by 21 climate models. The data can be viewed on a daily timescale for individual cities and towns and may be used to conduct climate risk assessments to predict the local and global effects of weather dangers, for example droughts, floods, heat waves and declines in agriculture productivity, and help plan responses to global warming effects..

REFERENCES

1. Joint-statement by leaders of 18 scientific organizations: American Association for the Advancement of Science, American Chemical Society, American Geophysical Union, American Institute of Biological Sciences, American Meteorological Society, American Society of Agronomy, American Society of Plant Biologists, American Statistical Association, Association of Ecosystem Research Centers, Botanical Society of America, Crop Science Society of America, Ecological Society of America, Natural Science Collections, Alliance Organization of Biological Field Stations, Society for Industrial and Applied Mathematics, Society of Systematic Biologists, Soil Science Society of America, University Corporation for Atmospheric Research (October 21, 2009), Joint-statement on climate change by leaders of 18 scientific organizations (PDF), Washington DC, USA: American Association for the Advancement of Science . Archived 14 July 2014.
2. Cramer, W., et al., Executive summary, in: Chapter 18: Detection and attribution of observed impacts (archived 18 October 2014), pp.982-984, in IPCC AR5 WG2 A 2014

3. Settele, J., et al., Section 4.3.2.1: Phenology, in: Chapter 4: Terrestrial and inland water systems (archived 20 October 2014), p.291, in IPCC AR5 WG2 A 2014
4. Oppenheimer, M., et al., Section 19.7.1: Relationship between Adaptation Efforts, Mitigation Efforts, and Residual Impacts, in: Chapter 19: Emergent risks and key vulnerabilities (archived 20 October 2014), pp.1080-1085, in IPCC AR5 WG2 A 2014
5. Oppenheimer, M., et al., Section 19.6.2.2. The Role of Adaptation and Alternative Development Pathways, in: Chapter 19: Emergent risks and key vulnerabilities (archived 20 October 2014), pp.1072-1073, in IPCC AR5 WG2 A 2014
6. Denton, F., et al., Section 20.3. Contributions to Resilience through Climate Change Responses, in: Chapter Climate-resilient pathways: adaptation, mitigation, and sustainable development (archived 20 October 2014), pp.1113-1118, in IPCC AR5 WG2 A 2014
7. Field, C.B., et al., Section A-3. The Decision-making Context, in: Technical summary (archived 18 October 2014), p.55, in IPCC AR5 WG2 A 2014
8. SPM.4.1 Long-term mitigation pathways, in: Summary for Policymakers, pp.11-15 (archived 2 July 2014), in IPCC AR5 WG3 2014
9. Clarke, L., et al., Section 6.3.1.3 Baseline emissions projections from fossil fuels and industry (pp.17-18 of final draft), in: Chapter 6: Assessing Transformation Pathways (archived 20 October 2014), in: IPCC AR5 WG3 2014
10. Greenhouse Gas Concentrations and Climate Implications, p.14, in Prinn & Reilly 2014. The range given by Prinn and Reilly is 3.3 to 5.5 °C, with a median of 3.9 °C.
11. SPM.3 Trends in stocks and flows of greenhouse gases and their drivers, in: Summary for Policymakers, p.8 (archived 2 July 2014), in IPCC AR5 WG3 2014. The range given by the Intergovernmental Panel on Climate Change is 3.7 to 4.8 °C, relative to pre-industrial levels (2.5 to 7.8 °C including climate uncertainty).
12. Field, C.B., et al., Box TS.8: Adaptation Limits and Transformation, in: Technical summary (archived 18 October 2014), p.89, in IPCC AR5 WG2 A 2014
13. Field, C.B., et al., Section B-1. Key Risks across Sectors and Regions, in: Technical summary (archived 18 October 2014), p.62, in IPCC AR5 WG2 A 2014
14. IPCC, Glossary A-D "climate" and "climate change", in IPCC AR4 WG1 2007
15. US Environmental Protection Agency (US EPA) (14 June 2012), Climate Change Science Overview, US EPA, Click on the image to open a pop-up that explains the differences between climate change and global warming.
16. Albritton; et al., "Technical Summary", Box 1: What drives changes in climate?, , in IPCC TAR WG1 2001
17. Pielke, Roger; Gwyn Prins; Steve Rayner; Daniel Sarewitz (2007-02-08). "Lifting the taboo on adaptation" (PDF). Nature. 445: 597–8. doi:10.1038/445597a. Retrieved 2012-08-09.
18. "Definition of Anthropogenic Forcing – Chemistry Dictionary and Glossary".

19. Herring, D. (March 6, 2012). "ClimateWatch Magazine " Global Temperature Projections". NOAA Climate Portal.
20. Schneider; et al., "Chapter 19: Assessing key vulnerabilities and the risk from climate change", Sec. 19.3.1 Introduction to Table 19.1 Missing or empty |title= (help), in IPCC AR4 WG2 2007.
21. IPCC, "Summary for Policymakers", Sec. 1. Observed changes in climate and their effects Missing or empty |title= (help), p. 5, in IPCC AR4 SYR 2007.
22. Fisher; et al., "Chapter 3: Issues related to mitigation in the long-term context", Sec. 3.1 Emissions scenarios Missing or empty |title= (help), in IPCC AR4 WG3 2007
23. Karl 2009 (eds.). "Global Climate Change". (PDF)http://downloads.globalchange.gov/usimpacts/pdfs/climate-impacts-report.pdf. Missing or empty |title= (help)
24. Morita; et al., "Chapter 2, Greenhouse Gas Emission Mitigation Scenarios and Implications", Sec. 2.5.1.1 IPCC Emissions Scenarios and the SRES Process Missing or empty |title= (help), in IPCC TAR WG3 2001.
25. IPCC, "Synthesis report", Sec. 3.1 Emissions scenarios Missing or empty |title= (help), in IPCC AR4 SYR 2007.
26. IPCC, "Summary for Policymakers", Sec. 1 3. Projected climate change and its impacts: Table SPM.1 Missing or empty |title= (help), in IPCC AR4 SYR 2007.
27. Morita; et al., "Chapter 2, Greenhouse Gas Emission Mitigation Scenarios and Implications", Sec. 2.5.1.4 Emissions and Other Results of the SRES Scenarios Missing or empty |title= (help), in IPCC TAR WG3 2001.
28. Temperature change in years 2090-2099 relative to years 1980-1999
29. Solomon; et al., "Technical Summary", Box TS.1: Treatment of Uncertainties in the Working Group I AssessmentMissing or empty |title= (help), in IPCC AR4 WG1 2007.
30. United Nations Environment Programme (UNEP) (November 2010), "Ch 2: Which emissions pathways are consistent with a 2 °C or a 1.5 °C temperature limit?: Sec 2.2 What determines long-term temperature?", The Emissions Gap Report: Are the Copenhagen Accord pledges sufficient to limit global warming to 2 °C or 1.5 °C? A preliminary assessment (advance copy) (PDF), UNEP, p.28. This publication is also available in e-book format
31. "Box 8.1 Likelihood of exceeding a temperature increase at equilibrium, in: Ch 8: The Challenge of Stabilisation" (PDF), Missing or empty |title= (help), in Stern 2006, p. 195
32. Solomon, S.; et al. (January 28, 2009). "Irreversible climate change due to carbon dioxide emissions" (PDF). Proceedings of the National Academy of Sciences of the United States of America. US National Academy of Sciences. 106 (6): 1704–9. Bibcode:2009PNAS..106.1704S. doi:10.1073/pnas.0812721106. PMC 2632717 . PMID 19179281.
33. "Question 5", Figure 5-2 Missing or empty |title= (help), in IPCC TAR SYR 2001, p. 89

34. Meehl, G.A.; et al., "Ch 10: Global Climate Projections", Sec 10.7.2 Climate Change Commitment to Year 3000 and Beyond to EquilibriumMissing or empty |title= (help), in IPCC AR4 WG1 2007
35. Hansen, J.E.; M. Sato (July 2011), NASA GISS: Science Briefs: Earth's Climate History: Implications for Tomorrow, New York City, New York, USA: NASA GISS
36. Overpeck, J.T. (20 August 2008), NOAA Paleoclimatology Global Warming - The Story: Proxy Data, NOAA Paleoclimatology Program - NCDC Paleoclimatology Branch
37. Committee on Surface Temperature Reconstructions for the Last 2,000 Years, US National Research Council (2006), "Summary", Surface Temperature Reconstructions for the Last 2,000 Years, Washington, D.C., USA: National Academy Press, ISBN 0-309-66144-7 External link in |publisher= (help) , p.3
38. Jansen; et al., "Chapter 6: Palaeoclimate", Sec. 6.3.2 What Does the Record of the Mid-Pliocene Show? Missing or empty |title=(help), in IPCC AR4 WG1 2007.
39. Smith, J.B.; et al., "Ch 19. Vulnerability to Climate Change and Reasons for Concern: A Synthesis", Footnote 4, in Sec 19.8.2. What does Each Reason for Concern Indicate? Missing or empty |title=(help), in IPCC TAR WG2 2001, p. 957
40. NOAA 2010, p. 2
41. NOAA 2010, p. 3
42. Solomon; et al., "Technical Summary", TS.3.4 Consistency Among Observations Missing or empty |title= (help), in IPCC AR4 WG1 2007.
43. Rosenzweig; et al., "Chapter 1: Assessment of Observed Changes and Responses in Natural and Managed Systems", Executive summary Missing or empty |title= (hclp), in IPCC AR4 WG2 2007.
44. Hegerl; et al., "Chapter 9: Understanding and Attributing Climate Change", Executive Summary Missing or empty |title= (help), in IPCC AR4 WG1 2007.
45. IPCC, "Summary for Policymakers", Human and Natural Drivers of Climate Change Missing or empty |title= (help), in IPCC AR4 WG1 2007.
46. Committee on the Science of Climate Change, US National Research Council (2001). "3. Human Caused Forcings". Climate Change Science: An Analysis of Some Key Questions. Washington, D.C., USA: National Academy Press. p. 12. ISBN 0-309-07574-2. External link in |publisher= (help)
47. IPCC, "Summary for Policymakers", Sec. 3. Projected climate change and its impacts Missing or empty |title= (help), in IPCC AR4 SYR 2007.
48. ESRL web team (26 January 2009). "ESRL News: New Study Shows Climate Change Largely Irreversible" (Press release). US Department of Commerce, NOAA, Earth System Research Laboratory (ESRL).
49. IPCC, "Summary for Policymakers", Magnitudes of impactMissing or empty |title= (help), p.17, IPCC AR4 WG2 2007.
50. Executive Summary (PHP). Abrupt Climate Change: Inevitable Surprises. United States National Academy of Sciences. June 2002.

51. Le Treut; et al., "Chapter 1: Historical Overview of Climate Change Science", FAQ 1.2 What is the Relationship between Climate Change and Weather? Missing or empty |title= (help), in IPCC AR4 WG1 2007.
52. NOAA Geophysical Fluid Dynamics Laboratory (GFDL) (9 October 2012), GFDL Climate Modeling Research Highlights: Will the Wet Get Wetter and the Dry Drier, NOAA GFDL
53. NOAA (February 2007), "Will the wet get wetter and the dry drier?" (PDF), GFDL Climate Modeling Research Highlights, Princeton, NJ, USA: National Oceanic and Atmospheric Administration (NOAA) Geophysical Fluid Dynamics Laboratory (GFDL), 1 (5). Revision 10/15/2008, 4:47:16 PM.
54. "Summary for policymakers", https://docs.google.com/file/d/0B1gFp6Ioo3akYklZcWkwWHJud00/edit?pli=1 Missing or empty |title= (help), in IPCC SREX 2012, p. 8
55. IPCC (2013), Table SPM.1, in Summary for Policymakers, p. 5 (archived PDF), in IPCC AR5 WG1 2013
56. Stocker, T.F., et al. (2013), Temperature Extremes, Heat Waves and Warm Spells, in: TFE.9, in: Technical Summary, p. 111 (archived PDF), in IPCC AR5 WG1 2013
57. Stocker, T.F., et al. (2013), Floods and Droughts, in: TFE.9, in: Technical Summary, p. 112 (archived PDF), in IPCC AR5 WG1 2013
58. Christensen, J.H.,et al. (2013), Cyclones, in: Executive Summary, in: Chapter 14: Climate Phenomena and their Relevance for Future Regional Climate Change, p. 1220 (archived PDF), in IPCC AR5 WG1 2013
59. Confalonieri; et al., "Chapter 8: Human health", Executive summary Missing or empty |title= (help), in IPCC AR4 WG2 2007.
60. IPCC, "Summary for Policymakers", Question 4 Missing or empty |title= (help), p.14, in IPCC TAR SYR 2001.
61. IPCC, "Synthesis Report, Topic 5: The long-term perspective", Sec. 5.2 Key vulnerabilities, impacts and risks – long-term perspectives Missing or empty |title= (help), pp. 64–65, in IPCC AR4 SYR 2007.
62. "Cryosphere", in US EPA 2012
63. 2011 Arctic Sea Ice Minimum, in Kennedy 2012
64. Mass Balance of Mountain Glaciers in 2011, in Kennedy 2012
65. 2011 Snow Cover in Northern Hemisphere, in Kennedy 2012
66. Solomon, S.; et al., "Technical summary", TS.5.2 Large-Scale Projections for the 21st Century Missing or empty |title= (help), in IPCC AR4 WG1 2007
67. Met Office, Arctic sea ice 2012, Exeter, UK: Met Office
68. Meehl, G.A.; et al., "Ch 10: Global Climate Projections", Box 10.1: Future Abrupt Climate Change, 'Climate Surprises', and Irreversible Changes: Glaciers and ice caps Missing or empty |title= (help), in IPCC AR4 WG1 2007, p. 776
69. Meehl, G.A.; et al., "Ch 10: Global Climate Projections", Sec 10.3.3.2 Changes

in Snow Cover and Frozen Ground Missing or empty |title= (help), in IPCC AR4 WG1 2007, pp. 770, 772

70. Field, C.B.; et al., "Ch 14: North America", Sec 14.4.1 Freshwater resources: Surface water Missing or empty |title= (help), in IPCC AR4 WG2 2007, p. 627
71. "Snowpack", in US EPA 2012
72. Some of these impacts are included in table SPM.2: "Summary for Policymakers", 3 Projected climate change and its impacts: Table SPM.2 Missing or empty |title= (help), in IPCC AR4 SYR 2007, pp. 11–12
73. "Ch 3: Fresh Water Resources and their Management", Sec 3.4.3 Floods and droughts Missing or empty |title= (help), in IPCC AR4 WG2 2007, p. 187
74. Barnett, T.P.; et al. (17 November 2005), "Potential impacts of a warming climate on water availability in snow-dominated regions: Abstract" (PDF), Nature, 438, pp. 303–9, Bibcode:2005Natur.438..303B, doi:10.1038/nature04141, PMID 16292301
75. State of the Climate in 2009, as appearing in the July 2010 issue (Vol. 91) of the Bulletin of the American Meteorological Society (BAMS). Supplemental and Summary Materials: Report at a Glance: Highlights. Website of the US National Oceanic and Atmospheric Administration: National Climatic Data Center. July 2010. Retrieved 2011-06-06.
76. Bindoff, N.L.; et al., "Ch. 5: Observations: Oceanic Climate Change and Sea Level", Sec 5.2.2.3 Implications for Earth's Heat BalanceMissing or empty |title= (help), in IPCC AR4 WG1 2007, referred to by: Climate Graphics by Skeptical Science: Global Warming Components:, Skeptical Science, Components of global warming for the period 1993 to 2003 calculated from IPCC AR4 5.2.2.3
77. This article incorporates public domain material from the US EPA document: Climate Change Indicators in the United States, 2012, 2nd ed: Ocean Acidity: Figure 2. Changes in Aragonite Saturation of the World's Oceans, 1880-2012, US Environmental Protection Agency (EPA)
78. "Introduction". Ecological Impacts of Climate Change. p. 8., in NRC 2008b
79. IPCC, "Synthesis Report", Sec. 3.3.4 Ocean acidificationMissing or empty |title= (help), in IPCC AR4 SYR 2007.
80. Fischlin; et al., "Chapter 4: Ecosystems, their properties, goods, and services", Executive summary [??] Missing or empty |title= (help), p. 213, in IPCC AR4 WG2 2007.
81. Crowley, T. J.; North, G. R. (May 1988). "Abrupt Climate Change and Extinction Events in Earth History". Science. 240 (4855): 996–1002. Bibcode:1988Sci...240..996C. doi:10.1126/science.240.4855.996. PMID 17731712.
82. Shaffer, G. .; Olsen, S. M.; Pedersen, J. O. P. (2009). "Long-term ocean oxygen depletion in response to carbon dioxide emissions from fossil fuels". Nature Geoscience. 2 (2): 105–109. Bibcode:2009NatGe...2..105S. doi:10.1038/ngeo420.
83. US Environmental Protection Agency (US EPA) (2010). "Sea Level: Climate Change: US EPA". US EPA.

84. Bindoff; et al., "Chapter 5: Observations: Oceanic Climate Change and Sea Level", FAQ 5.1 Is Sea Level Rising? Missing or empty |title= (help), in IPCC AR4 WG1 2007.
85. Bindoff; et al., "Chapter 5: Observations: Oceanic Climate Change and Sea Level", Executive summary Missing or empty |title=(help), in IPCC AR4 WG1 2007.
86. Albritton; et al., "Technical Summary", Box 2: What causes sea level to change? Missing or empty |title= (help), in IPCC TAR WG1 2007.
87. "7 Sea Level Rise and the Coastal Environment", http://books.nap.edu/openbook.php?record_id=12782&page=245Missing or empty |title= (help), in NRC 2010, p. 245.
88. IPCC, "Section 3: Projected climate change and its impacts", Table SPM.1 Missing or empty |title= (help), in IPCC AR4 SYR 2007.
89. change in sea level by 2090–99, relative to 1980–99
90. Meehl; et al., "Chapter 10: Global Climate Projections", Executive summary Missing or empty |title= (help), in IPCC AR4 WG1 2007.
91. "7 Sea Level Rise and the Coastal Environment", pp. 243–245 http://books.nap.edu/openbook.php?record_id=12782&page=243Missing or empty |title= (help), in US NRC 2010.
92. Bindoff; et al., "Chapter 5: Observations: Oceanic Climate Change and Sea Level", Missing or empty |title= (help), in IPCC AR4 WG1 2007.
93. Gille, Sarah T. (February 15, 2002). "Warming of the Southern Ocean Since the 1950s". Science. 295 (5558): 1275–7. Bibcode:2002Sci...295.1275G. doi:10.1126/science.1065863. PMID 11847337.
94. US Environmental Protection Agency (14 June 2012). "Science: Climate Change: US EPA (Climate Change Science Overview)". US EPA.
95. Rosenzweig, C. (December 2008). "Science Briefs: Warming Climate is Changing Life on Global Scale". Website of the US National Aeronautics and Space Administration, Goddard Institute for Space Studies. Retrieved 2011-07-08.
96. US NRC (2008). Understanding and Responding to Climate Change. A brochure prepared by the US National Research Council (US NRC) (PDF). Washington DC, USA: Board on Atmospheric Sciences and Climate, National Academy of Sciences. External link in |publisher= (help)
97. IPCC, "Summary for Policymakers", Projections of Future Changes in Climate Missing or empty |title= (help), in IPCC AR4 WG1 2007.
98. IPCC, "Synthesis Report", Question 9: Table SPM-3 Missing or empty |title= (help), in IPCC TAR SYR 2001.
99. IPCC, "Synthesis report", Sec. 3.3.3 Especially affected systems, sectors and regions Missing or empty |title= (help), in IPCC AR4 SYR 2007.
100. Schneider, S.H.; et al., "Ch 19: Assessing Key Vulnerabilities and the Risk from Climate Change", Distribution of Impacts, in: Sec 19.3.7 Update on 'Reasons for Concern' Missing or empty |title= (help), in IPCC AR4 WG2 2007, p. 796

101. Schneider, S.H.; et al., "Ch 19: Assessing Key Vulnerabilities and the Risk from Climate Change", Sec 19.3.3 Regional vulnerabilitiesMissing or empty |title= (help), in IPCC AR4 WG2 2007, p. 792
102. Wilbanks, T.J.; et al., "Ch 7: Industry, Settlement and Society", Sec 7.4.2.5 Social issues and Sec 7.4.3 Key vulnerabilities Missing or empty |title= (help), in IPCC AR4 WG2 2007, pp. 373–376
103. IPCC, Glossary P-Z: "sensitivity" and "vulnerability", in IPCC AR4 WG2 2007.
104. IPCC, "Summary for Policymakers", Sec. 2.4. Many Human Systems are Sensitive to Climate Change, and Some are VulnerableMissing or empty |title= (help), in IPCC TAR WG2 2001.
105. Schneider; et al., "Chapter 19: Assessing key vulnerabilities and the risk from climate change", 19.3.2.2 Other market sectors Missing or empty |title= (help), in IPCC AR4 WG2 2007.
106. Easterling; et al., "Chapter 5: Food, Fibre, and Forest Products", 5.4.1 Primary effects and interactions Missing or empty |title=(help), in IPCC AR4 WG2 2007, p. 282.
107. Schneider; et al., "Chapter 19: Assessing Key Vulnerabilities and the Risk from Climate Change", Sec. 19.3.2.1 Agriculture Missing or empty |title= (help), in IPCC AR4 WG2 2007, p. 790.
108. Rosenzweig; et al., "Chapter 1: Assessment of Observed Changes and Responses in Natural and Managed Systems", Sec. 1.3.6.1 Crops and livestock Missing or empty |title= (help), in IPCC AR4 WG2 2007.
109. IPCC, Glossary P-Z: "phenology", in IPCC AR4 WG2 2007.
110. Figure 5.1, p.161, in: Sec 5.1 FOOD PRODUCTION, PRICES, AND HUNGER, in: Ch 5: Impacts in the Next Few Decades and Coming Centuries, in: US NRC 2011
111. Sec 5.1 FOOD PRODUCTION, PRICES, AND HUNGER, pp.160-162, in: Ch 5: Impacts in the Next Few Decades and Coming Centuries, in US NRC 2011
112. Schneider, S.H.; et al., "Chapter 19: Assessing Key Vulnerabilities and the Risk from Climate Change", Box 19.2. Reference for temperature levels Missing or empty |title= (help), in IPCC AR4 WG2 2007, p. 783.
113. Easterling; et al., "Chapter 5: Food, Fibre, and Forest Products", Sec. 5.6.5 Food security and vulnerability Missing or empty |title=(help), in IPCC AR4 WG2 2007.
114. "World hunger increasing". Food and Agriculture Organization (FAO) Newsroom. 30 October 2006. Retrieved 2011-07-07.
115. Dai, A. (2011). "Drought under global warming: A review". Wiley Interdisciplinary Reviews: Climate Change. 2: 45–65. doi:10.1002/wcc.81.
116. Sheffield, J.; Wood, E. F.; Roderick, M. L. (2012). "Little change in global drought over the past 60 years". Nature. 491 (7424): 435–438. Bibcode:2012Natur.491..435S. doi:10.1038/nature11575.
117. Mishra, A. K.; Singh, V. P. (2011). "Drought modeling – A review". Journal

of Hydrology. 403: 157–175. Bibcode:2011JHyd..403..157M. doi:10.1016/j.jhydrol.2011.03.049.

118. Ding, Y.; Hayes, M. J.; Widhalm, M. (2011). "Measuring economic impacts of drought: A review and discussion". Disaster Prevention and Management. 20 (4): 434–446. doi:10.1108/09653561111161752.

119. WHO (2009). "Ch. 2, Results: 2.6 Environmental risks" (PDF). Global health risks: mortality and burden of disease attributable to selected major risks (PDF). Geneva, Switzerland: WHO Press. p. 24. ISBN 978 92 4 156387 1.

120. IPCC AR4 SYR 2007.

121. Confalonieri; et al., "Chapter 8: Human health", Sec. 8.3.2 Future vulnerability to climate change Missing or empty |title= (help), in IPCC AR4 WG2 2007.

122. Kundzewicz; et al., "Chapter 3: Fresh Water Resources and their Management", Sec. 3.2 Current sensitivity/vulnerability Missing or empty |title= (help), in IPCC AR4 WG2 2007.

123. Kundzewicz; et al., "Chapter 3: Fresh Water Resources and their Management", Sec. 3.3 Executive summary Missing or empty |title= (help), p. 175, in IPCC AR4 WG2 2007.

124. Scott; et al., "Chapter 12: Human settlements in a changing climate: impacts and adaptation", Sec. 12.3.1 Population MigrationMissing or empty |title= (help), pp. 406—407, in IPCC SAR WG2 1996.

125. Special Report on Emissions Scenarios, 10.2.5.1. Human Settlements, https://www.ipcc.ch/ipccreports/sres/regional/275.htm

126. The World Bank, "Part One: Chapter 2: Reducing Human Vulnerability: Helping People Help Themselves" (PDF), Managing social risks: Empower communities to protect themselves, p. 109, WDR 2010.

127. Desanker; et al., "Chapter 10: Africa", Executive summaryMissing or empty |title= (help), in IPCC TAR WG2 2001.

128. Scott; et al., "Chapter 7: Human Settlements, Energy, and Industry", Sec. 7.2.2.3.1. Migration Missing or empty |title=(help), in IPCC TAR WG2 2001.

129. Wilbanks; et al., "Chapter 7: Industry, settlement and society", Sec. 7.4.1 General effects: Box 7.2. Environmental migration Missing or empty |title= (help), in IPCC AR4 WG2 2007.

130. Hsiang SM, Burke M, Miguel E (September 2013). "Quantifying the influence of climate on human conflict". Science. 341 (6151): 1235367. doi:10.1126/science.1235367. PMID 24031020.

131. Ranson, M. (2014). "Crime, weather, and climate change". Journal of Environmental Economics and Management. 67: 274–302. doi:10.1016/j.jeem.2013.11.008.

132. Spaner, J S; LeBali, H (October 2013). "The Next Security Frontier". Proceedings of the US Naval Institute. 139 (10): 30–35. Retrieved 23 Nov 2015.

133. Glossary A-D: "Aggregate impacts", in IPCC AR4 SYR 2007, p. 76

134. Smith; et al., "Chapter 19: Vulnerability to Climate Change and Reasons for Concern: A Synthesis", Sec. 19.5.3 Insights and Lessons: Vulnerability over Time: Table 19-5 Missing or empty |title= (help), in IPCC TAR WG2 2001.

135. Smith; et al., "Chapter 19: Vulnerability to Climate Change and Reasons for Concern: A Synthesis", Sec. 19.5.2. Insights and Lessons: The Static Picture Missing or empty |title= (help), in IPCC TAR WG2 2001.

136. Rosenzweig; et al., "Chapter 1: Assessment of observed changes and responses in natural and managed systems", Sec. 1.3.8.5 Summary of disasters and hazards Missing or empty |title= (help), in IPCC AR4 WG2 2007.

137. IPCC, "Synthesis Report", Question 2, Sections 2.25 and 2.26 Missing or empty |title= (help), p. 55, IPCC TAR SYR 2001.

138. Billion-Dollar Weather and Climate Disasters

139. Schneider; et al., "Chapter 19: Assessing key vulnerabilities and the risk from climate change", Sec. 19.3.2.3 Aggregate market impactsMissing or empty |title= (help), p. 790, in IPCC AR4 WG2 2007.

140. Smith; et al., "Chapter 19: Vulnerability to Climate Change and Reasons for Concern: A Synthesis", Sec. 19.8.2.3. Aggregate Impacts Missing or empty |title= (help), in IPCC TAR WG2 2001.

141. Schneider; et al., "Chapter 19: Assessing key vulnerabilities and the risk from climate change", Sec. 19.3.7 Update on 'Reasons for Concern' Missing or empty |title= (help), p. 796, in IPCC AR4 WG2 2007.

142. Smith; et al., "Chapter 19: Vulnerability to Climate Change and Reasons for Concern: A Synthesis", Sec. 19.4.3. Distribution of Total Impacts Missing or empty |title= (help), in IPCC TAR WG2 2001.

143. NRC 2008b. "Introduction". Ecological Impacts of Climate Change. p. 14.

144. NRC 2008b. "Introduction". Ecological Impacts of Climate Change. p. 16.

145. Fischlin; et al., "Chapter 4: Ecosystems, their properties, goods and services", Executive summary Missing or empty |title=(help), pp. 213–214, in IPCC AR4 WG2 2007.

146. IPCC, "Synthesis Report, Question 3", Sec. 3.18http://www.grida.no/climate/ipcc_tar/vol4/english/026.htm Missing or empty |title= (help), in IPCC TAR SYR 2001.

147. Van Riper, Charles. (2014) Projecting Climate Effects on Birds and Reptiles of the Southwestern United States. Reston, Va.: U.S. Department of the Interior, U.S. Geological Survey.

148. Fischlin; et al., "Chapter 4: Ecosystems, their properties, goods and services", Sec. 4.4.11 Global synthesis including impacts on biodiversity Missing or empty |title= (help), in IPCC AR4 WG2 2007.

149. IPCC, Glossary A-D: "biota", in IPCC AR4 WG2 2007.

150. NRC, "Introduction", Ecological Impacts of Climate Change, p. 16, in NRC 2008b.

151. IPCC, "Section 3: Projected climate change and its impacts", Sec. 3.3 Impacts of future climate changes Missing or empty |title=(help), in IPCC AR4 SYR 2007.

152. IPCC, "Section 3: Projected climate change and its impacts", Sec. 3.3.1 Impacts on systems and sectors: Ecosystems Missing or empty |title= (help), in IPCC AR4 SYR 2007.

153. Smith; et al., "Chapter 19: Vulnerability to Climate Change and Reasons for Concern: A Synthesis", Sec. 19.8.2.5. Large-Scale Singularities Missing or empty |title= (help), in IPCC TAR WG2 2001.

154. White; et al., "Technical summary", Sec. 7.2.5, Large-Scale Singular Events Missing or empty |title= (help), in IPCC TAR WG2 2001.

155. "[F]or the most part" refers to improved scientific understanding of singularities since the assessment by White et al. (2001). See: Schneider; et al., "Chapter 19: Assessing key vulnerabilities and the risk from climate change", Sec. 19.3.7 Update on 'Reasons for Concern' Missing or empty |title= (help), in IPCC AR4 WG2 2007.

156. Schneider; et al., "Chapter 19: Assessing key vulnerabilities and the risk from climate change", Sec. ??? Missing or empty |title=(help), in IPCC AR4 WG2 2007.

157. Riebeek, H.. design by R. Simmon (May 9, 2006). "Paleoclimatology: Explaining the Evidence: Explaining Rapid Climate Change: Tales from the Ice". NASA Earth Observatory. Retrieved 2011-10-16.

158. CCSP (2008b). Abrupt Climate Change. A report by the US Climate Change Science Program (CCSP) and the Subcommittee on Global Change Research. Reston, VA: US Geological Survey.

159. Schneider; et al., "Chapter 19: Assessing key vulnerabilities and the risk from climate change", Sec. 19.3.5.3 Possible changes in the North Atlantic meridional overturning circulation (MOC) Missing or empty |title= (help), in IPCC AR4 WG2 2007.

160. Albritton; et al., "Technical Summary", Sec. C.1 Observed Changes in Globally Well-Mixed Greenhouse Gas Concentrations and Radiative Forcing Missing or empty |title= (help), p. 38, in IPCC TAR WG1 2001.

161. IPCC, "Summary for Policymakers", Sec. 2. Causes of changeMissing or empty |title= (help), p. 5, in IPCC AR4 SYR 2007.

162. Meehl; et al. "Chapter 10: Global Climate Projections". Frequently Asked Question 10.3: If Emissions of Greenhouse Gases are Reduced, How Quickly do Their Concentrations in the Atmosphere Decrease?.Missing or empty |title= (help), in IPCC AR4 WG1 2007.

163. Barker; et al., "Chapter 11: Mitigation from a cross-sectoral perspective", Executive summary - Unconventional options Missing or empty |title= (help) , in IPCC AR4 WG3 2007.

164. IPCC, "5.14", Question 5 http://www.grida.no/climate/ipcc_tar/vol4/english/044.htm Missing or empty |title= (help), in IPCC TAR SYR 2001.

165. Schneider; et al., “Chapter 19: Assessing key vulnerabilities and the risk from climate change”, Sec.19.2 Criteria for selecting ‘key’ vulnerabilities: Persistence and reversibility Missing or empty |title=(help), in IPCC AR4 WG2 2007.
166. Barnett, J; WN Adger (2003). “Climate dangers and atoll countries”(PDF). Climatic Change. Kluwer Academic Publishers. 61: 321–337. doi:10.1023/b:clim.0000004559.08755.88. This paper was published in 2001 as Tyndall Centre Working Paper 9
167. Tol Correction
168. Mendelsohn 2006,
169. Increased growing season in Greenland, Nyegaard 2007
170. Kimball 2007
171. Variations in northern vegetation activity inferred from satellite data of vegetation index during 1981 to 1999 L. Zhou et. al. 2001
172. Impact of climate forcing on ecosystem processes in the North Pacific Subtropical Gyre Guido Corno et. al. 2007
173. McMahon 2010
174. Klady 2010
175. Climate warming will not decrease winter mortalityP.L. Staddon et.al. 2014
176. Kerr 2002
177. Ducklow 2006
178. Ozgul 2010
179. IPCC 2010, pp. 4–7
180. IPCC press release 2014
181. Molina & others n.d.
182. PBL & others 2009
183. For example: Good & others (2010). Refer to UKMO (2013) for other AVOID publications.
184. UK Royal Society & US National Academy of Sciences 2014
185. Allison & others 2009
186. US NRC 2010
187. US NRC 2011
188. Molina & others n.d., p. 3
189. “NASA Releases Detailed Global Climate Change Projections”. Retrieved 2015-06-09.
190. IPCC, “Synthesis Report”, Question 2, Box 2-1 Missing or empty |title= (help), IPCC TAR SYR 2001.
191. Moss, R.; Schneider, S. (July 2000), “Uncertainties”, in R. Pachuari; T. Taniguchi; K. Tanaka (eds.), IPCC supporting material: guidance papers on the cross cutting issues of the Third Assessment Report of the IPCC (PDF), 2-1-1 Toranomon,

Minato-ku, 1050001 Tokyo, Japan: Global Industrial and Social Progress Research Institute (GISPRI), ISBN 4-9980908-0-1

192. White; et al., "Technical Summary", Box 2. Confidence Levels and State of Knowledge Missing or empty |title= (help), in IPCC TAR WG2 2001.
193. "Summary for policymakers", https://docs.google.com/file/d/0B1gFp6Ioo3akYklZcWkwWHJud00/edit?pli=1 Missing or empty |title= (help), in IPCC SREX 2012, p. 19
194. Ahmad; et al., "Chapter 2: Methods and Tools", Sec. 2.6.2. "Objective" and "Subjective" Probabilities are not Always Explicitly Distinguished Missing or empty |title= (help), in IPCC TAR WG2 2001.
195. Granger Morgan, M. (Lead Author), H. Dowlatabadi, M. Henrion, D. Keith, R. Lempert, S. McBride, M. Small and T. Wilbanks (Contributing Authors) (2009). Synthesis and Assessment Product 5.2: Best practice approaches for characterizing, communicating, and incorporating scientific uncertainty in decisionmaking. A Report by the U.S. Climate Change Science Program and the Subcommittee on Global Change Research. Washington D.C., USA.: National Oceanic and Atmospheric Administration. pp. 19–20; 27–28. Tables 2.1 and 2.2 on pages 27-28 show two quantitative scales of uncertainty which are used in the TAR and AR4.

CHAPTER 9

Effects of Climate Change on Plant Biodiversity

Chapter Contents

- **Introduction**
- **Palaeo Context**
- **Modern Context**
- **Direct Impacts of Climate Change**
- **Indirect Impacts of Climate Change**
- **Higher Level Changes**
- **Challenges of Modelling Future Impacts**

Climate change has had a profound impact on current biogeography, so we can expect our ongoing and rapid climate change, to have a great an effect. Climate change has important implications for nearly every aspect of life on Earth, and consequences are already being felt.

Climate change is already having an impact on biodiversity, and is projected to become a progressively more significant threat in the coming decades. Loss of Arctic sea ice threatens biodiversity across an entire biome and beyond. The related pressure of ocean acidification, resulting from higher concentrations of carbon dioxide in the atmosphere, is also already being observed.

INTRODUCTION

Environmental conditions play a key role in defining the function and distribution of plants, in combination with other factors. Changes in long term environmental conditions that can be collectively coined climate change are known to have had enormous impacts on plant diversity patterns in the future and are seen as having significant current impacts. It is predicted that climate change will remain one of the major drivers of biodiversity patterns in the future.

Alpine flora at Logan Pass, Glacier National Park, in Montana, United States: Alpine plants are one group expected to be highly susceptible to the impacts of climate change

PALAEO CONTEXT

Australian Rainforest: An ecosystem known to have significantly contracted in area over recent geological time as a result of climatic changes.

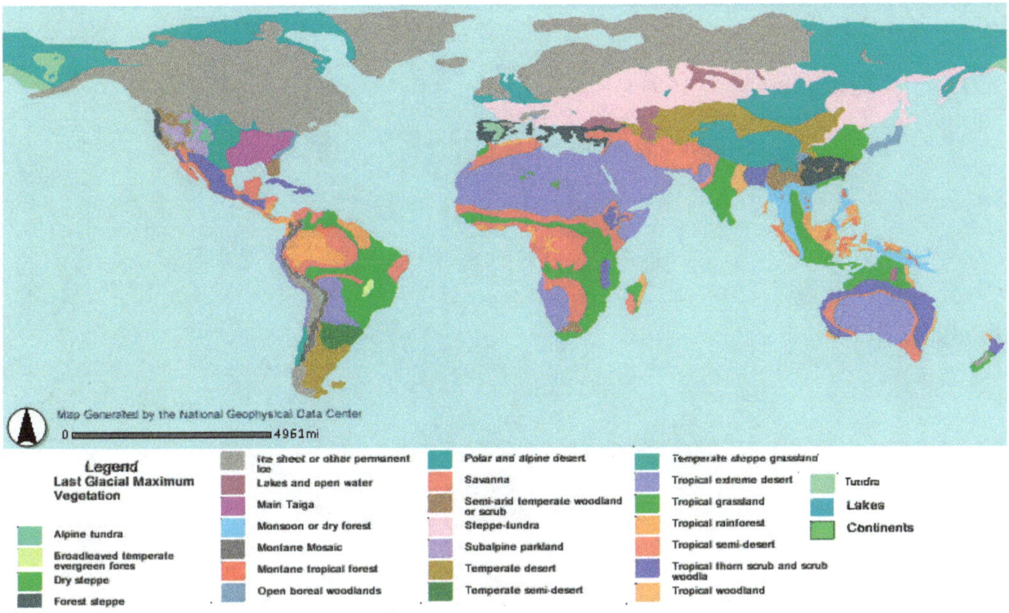

Map of global vegetation distributions during the last glacial maximum

The Earth has experienced a constantly changing climate in the time since plants first evolved. In comparison to the present day, this history has seen Earth as cooler, warmer, drier, and wetter, and CO_2 (carbon dioxide) concentrations have been both higher and lower. These changes have been reflected by constantly shifting vegetation, for example forest communities dominating most areas in interglacial periods, and herbaceous communities dominating during glacial periods. It has been shown that past climatic change has been a major driver of the processes of speciation and extinction. The best known example of this is the Carboniferous Rainforest Collapse which occurred 350 million years ago. This event decimated amphibian populations and spurred on the evolution of reptiles.

MODERN CONTEXT

There is significant current interest and research focus on the phenomenon of recent anthropogenic climate changes, or global warming. Focus is on identifying the current impacts of climate change on biodiversity, and predicting these effects into the future.

Changing climatic variables relevant to the function and distribution of plants include increasing CO_2 concentrations, increasing global temperatures, altered precipitation patterns, and changes in the pattern of 'extreme' weather events such as cyclones, fires, or storms.

Because individual plants and therefore species can only function physiologically, and successfully complete their life cycles under specific environmental conditions (ideally within a subset of these), changes to climate are likely to have significant impacts on plants from the level of the individual right through to the level of the ecosystem or biome.

Effects of CO_2

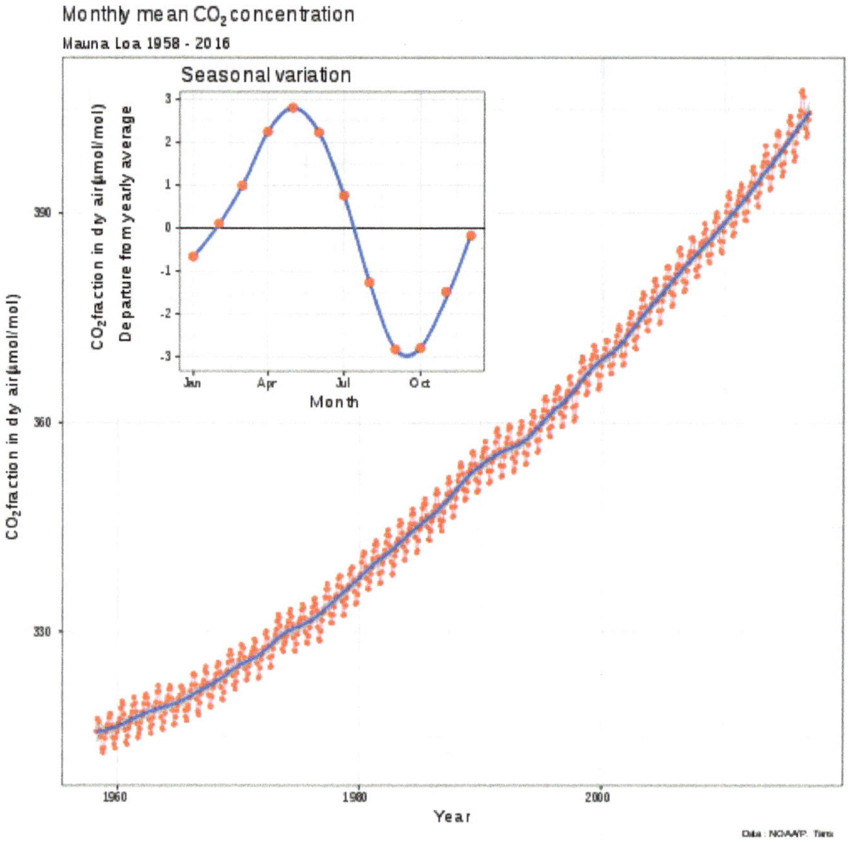

Recent increases in atmospheric CO_2.

Increases in atmospheric CO_2 concentration affect how plants photosynthesise, resulting in increases in plant water use efficiency, enhanced photosynthetic capacity, and increased growth. Increased CO_2 has been implicated in 'vegetation thickening' which affects plant community structure and function. Depending on the environment, there are differential responses to elevated atmospheric CO_2 between major 'functional types' of plant, such as C_3 and C_4 plants, or more or less woody species; which has the potential among other things to alter competition between these groups. Increased CO_2 can also lead to increased Carbon: Nitrogen ratios in the leaves of plants or in other aspects of leaf chemistry, possibly changing herbivore nutrition.

Effects of Temperature

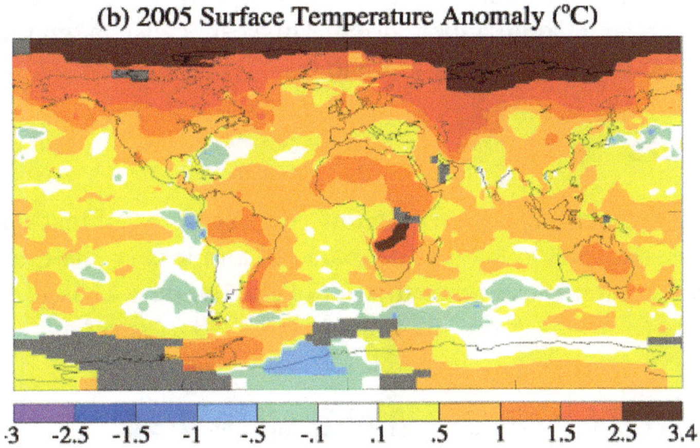

Global annual surface temperature anomaly in 2005, relative to 1951-1980 mean.

Increases in temperature raise the rate of many physiological processes such as photosynthesis in plants, to an upper limit. Extreme temperatures can be harmful when beyond the physiological limits of a plant.

Effects of Water

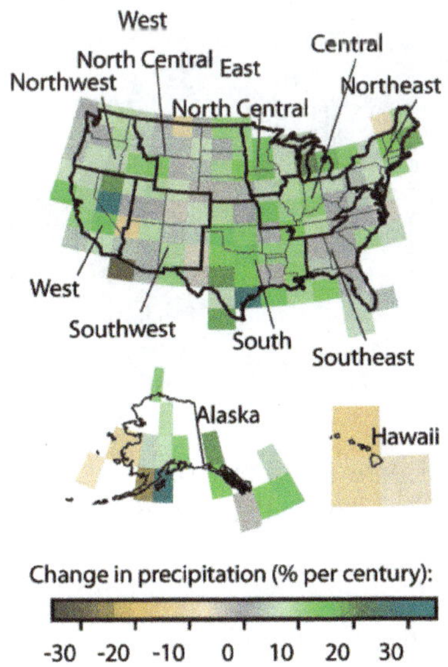

Precipitation trends in the United States, from the period 1901-2005. In some areas rainfall has increased in the last century, while some areas have dried.

As water supply is critical for plant growth, it plays a key role in determining the distribution of plants. Changes in precipitation are predicted to be less consistent than for temperature and more variable between regions, with predictions for some areas to become much wetter, and some much drier. This can cause a major change in some ecosystems which are dependent on water supply.

General Effects

Environmental variables will not act in isolation, but also in combination with one another, and with other pressures such as habitat degradation and loss or the introduction of exotic species. It is suggested that these other drivers of biodiversity change will act in synergy with climate change to increase the pressure on species to survive.

DIRECT IMPACTS OF CLIMATE CHANGE

Changes in Distributions

Pine tree representing an elevational tree-limit rise of 105 m over the period 1915–1974. Nipfjället, Sweden

If climatic factors such as temperature and precipitation change in a region beyond the tolerance of a species phenotypic plasticity, then distribution changes of the species may be inevitable. There is already evidence that plant species are shifting their ranges in altitude and latitude as a response to changing regional climates. Yet it is difficult to predict how species ranges will change in response to climate and separate these changes from all the other man-made environmental changes such as eutrophication, acid rain, and habitat destruction.

When compared to the reported past migration rates of plant species, the rapid pace of current change has the potential to not only alter species distributions, but also render many species as unable to follow the climate to which they are adapted. The environmental conditions required by some species, such as those in alpine regions may disappear altogether. The result of these changes is likely to be a rapid increase in extinction risk. Adaptation to new conditions may also be of great importance in the response of plants.

Predicting the extinction risk of plant species is not easy however. Estimations from particular periods of rapid climatic change in the past have shown relatively little species extinction in some regions, for example. Knowledge of how species may adapt or persist in the face of rapid change is still relatively limited.

Changes in the suitability of a habitat for a species drive distributional changes by not only changing the area that a species can physiologically tolerate, but how effectively it can compete with other plants within this area. Changes in community composition are therefore also an expected product of climate change.

Changes in Life-Cycles (Phenology)

The timing of phenological events such as flowering are often related to environmental variables such as temperature. Changing environments are therefore expected to lead to changes in life cycle events, and these have been recorded for many species of plants. These changes have the potential to lead to the asynchrony between species, or to change competition between plants. Flowering times in British plants for example have changed, leading to annual plants flowering earlier than perennials, and insect pollinated plants flowering earlier than wind pollinated plants; with potential ecological consequences. A recently published study has used data recorded by the writer and naturalist Henry David Thoreau to confirm effects of climate change on the phenology of some species in the area of Concord, Massachusetts.

INDIRECT IMPACTS OF CLIMATE CHANGE

All species are likely to be not only directly impacted by the changes in environmental conditions discussed above, but also indirectly through their interactions with other species. While direct impacts may be easier to predict and conceptualise, it is likely that indirect impacts are equally important in determining the response of plants to climate change. A species whose distribution changes as a direct result of climate change may 'invade' the range of another species or 'be invaded' for example, introducing a new competitive relationship or altering other processes such as carbon sequestration.

The range of a symbiotic fungi associated with plant roots may directly change as a result of altered climate, resulting in a change in the plant's distribution.

A new grass may spread into a region, altering the fire regime and greatly changing the species composition.

A pathogen or parasite may change its interactions with a plant, such as a pathogenic fungus becoming more common in an area where rainfall increases.

Increased temperatures may allow herbivores to expand further into alpine regions, significant impacting the composition of alpine herbfields.

There are innumerable examples of how climate change could indirectly affect plant species, most of which will be extremely difficult to predict.

HIGHER LEVEL CHANGES

Species respond in very different ways to climate change. Variation in the distribution, phenology and abundance of species will lead to inevitable changes in the relative abundance of species and their interactions. These changes will flow on to affect the structure and function of ecosystems.

CHALLENGES OF MODELLING FUTURE IMPACTS

Accurate predictions of the future impacts of climate change on plant diversity are critical to the development of conservation strategies. These predictions have come largely from bioinformatic strategies, involving modelling individual species, groups of species such as 'functional types', communities, ecosystems or biomes. They can also involve modelling species observed environmental niches, or observed physiological processes.

Although useful, modelling has many limitations. Firstly, there is uncertainty about the future levels of greenhouse gas emissions driving climate change and considerable uncertainty in modelling how this will affect other aspects of climate such as local rainfall or temperatures. For most species the importance of specific climatic variables in defining distribution (e.g. minimum rainfall or maximum temperature) is unknown. It is also difficult to know which aspects of a particular climatic variable are most biologically relevant, such as average vs. maximum or minimum temperatures. Ecological processes such as interactions between species and dispersal rates and distances are also inherently complex, further complicating predictions.

Improvement of models is an active area of research, with new models attempting to take factors such as life-history traits of species or processes such as migration into account when predicting distribution changes; though possible trade-offs between regional accuracy and generality are recognised.

Climate change is also predicted to interact with other drivers of biodiversity change such as habitat destruction and fragmentation, or the introduction of foreign species. These threats may possibly act in synergy to increase extinction risk from that seen in periods of rapid climate change in the past.

REFERENCES

1. Sahney, S., Benton, M.J. & Falcon-Lang, H.J. (2010). "Rainforest collapse triggered Pennsylvanian tetrapod diversification in Euramerica" (PDF). Geology. 38 (12): 1079–1082. doi:10.1130/G31182.1.
2. Dadamouny, M.A.; Schnittler, M. (2015). "Trends of climate with rapid change in Sinai, Egypt". J. of water and climate change. 7 (2). doi:10.2166/wcc.2015.215.

3. Sala OE, Chapin FS, Armesto JJ, et al. (March 2000). "Global biodiversity scenarios for the year 2100". Science. 287 (5459): 1770–4. doi:10.1126/science.287.5459.1770. PMID 10710299.
4. Duraiappah, Anantha K.; World Resources Institute (2006). Millennium Ecosystem Assessment: Ecosystems And Human-well Being—biodiversity Synthesis (PDF). Washington, D.C: World Resources Institute. ISBN 1-56973-588-3.
5. Dunlop, M., & Brown, P.R. (2008) Implications of climate change for Australia's National Reserve System: A preliminary assessment. Report to the Department of Climate Change, February 2008. Department of Climate Change, Canberra, Australia
6. Huntley, B. (2005). "North temperate responses". In Hannah, Lee Jay; Lovejoy, Thomas E. Climate Change and Biodiversity. New Haven, Conn: Yale University Press. pp. 109–24. ISBN 0-300-11980-1.
7. Steffen, W. & Canadell, P. (2005). 'Carbon Dioxide Fertilisation and Climate Change Policy.' 33 pp. Australian Greenhouse Office, Department of Environment and Heritage: Canberra
8. Gifford RM, Howden M (2001). "Vegetation thickening in an ecological perspective: significance to national greenhouse gas inventories". Environmental Science & Policy. 4: 59–72. doi:10.1016/S1462-9011(00)00109-X.
9. Jeffrey S. Dukes; Harold A. Mooney (April 1999). "Does global change increase the success of biological invaders?". Trends Ecol. Evol. (Amst.). 14 (4): 135–9. doi:10.1016/S0169-5347(98)01554-7. PMID 10322518.
10. Gleadow RM; et al. (1998). "Enhanced CO2 alters the relationship between photosynthesis and defence in cyanogenic Eucalyptus cladocalyx F. Muell.". Plant Cell Environ. 21: 12–22. doi:10.1046/j.1365-3040.1998.00258.x.
11. "National Climate Assessment". National Climate Assessment. Retrieved 2015-11-09.
12. Mackey, B. (2007). "Climate change, connectivity and biodiversity conservation". In Taylor M.; Figgis P. Protected Areas: buffering nature against climate change. Proceedings of a WWF and IUCN World Commission on Protected Areas symposium, Canberra, 18–19 June 2007. Sydney: WWF-Australia. pp. 90–6.
13. Lynch M.; Lande R. (1993). "Evolution and extinction in response to environmental change". In Huey, Raymond B.; Kareiva, Peter M.; Kingsolver, Joel G. Biotic Interactions and Global Change. Sunderland, Mass: Sinauer Associates. pp. 234–50. ISBN 0-87893-430-8.
14. Parmesan C, Yohe G (January 2003). "A globally coherent fingerprint of climate change impacts across natural systems". Nature. 421 (6918): 37–42. doi:10.1038/nature01286. PMID 12511946.
15. Walther GR, Post E, Convey P, et al. (March 2002). "Ecological responses to recent climate change". Nature. 416 (6879): 389–95. doi:10.1038/416389a. PMID 11919621.
16. Lenoir J, Gégout JC, Guisan A, Vittoz P, Wohlgemuth T, Zimmermann NE,

Dullinger S, Pauli H, Willner W, Svenning JC (2010). "Going against the flow: potential mechanisms for unexpected downslope range shifts in a warming climate". Ecography. 33: 295–303. doi:10.1111/j.1600-0587.2010.06279.x.

17. Groom, Q. (2012). "Some poleward movement of British native vascular plants is occurring, but the fingerprint of climate change is not evident". PeerJ. 1 (e77). doi:10.7717/peerj.77.
18. Hilbish TJ, Brannock PM, Jones KR, Smith AB, Bullock BN, Wethey DS (2010). "Historical changes in the distributions of invasive and endemic marine invertebrates are contrary to global warming predictions: the effects of decadal climate oscillations". Journal of Biogeography. 37: 423–431. doi:10.1111/j.1365-2699.2009.02218.x.
19. Davis MB, Shaw RG (April 2001). "Range shifts and adaptive responses to Quaternary climate change". Science. 292 (5517): 673–9. doi:10.1126/science.292.5517.673. PMID 11326089.
20. Thomas CD, Cameron A, Green RE, et al. (January 2004). "Extinction risk from climate change". Nature. 427 (6970): 145–8. doi:10.1038/nature02121. PMID 14712274.
21. Jump A, Penuelas J (2005). "Running to stand still: adaptation and the response of plants to rapid climate change". Ecol. Lett. 8: 1010–20. doi:10.1111/j.1461-0248.2005.00796.x.
22. Botkin DB; et al. (2007). "Forecasting the effects of global warming on biodiversity". BioScience. 57 (3): 227–36. doi:10.1641/B570306.
23. Fitter AH, Fitter RS (May 2002). "Rapid changes in flowering time in British plants". Science. 296 (5573): 1689–91. doi:10.1126/science.1071617. PMID 12040195.
24. Willis CG, Ruhfel B, Primack RB, Miller-Rushing AJ, Davis CC (November 2008). "Phylogenetic patterns of species loss in Thoreau's woods are driven by climate change". Proc. Natl. Acad. Sci. U.S.A. 105 (44): 17029–33. doi:10.1073/pnas.0806446105. PMC 2573948Freely accessible. PMID 18955707.
25. Dadamouny, M.A. (2009). "Population Ecology of Moringa peregrina growing in Southern Sinai, Egypt.". M.Sc. Suez Canal University, Faculty of Science, Botany Department. p. 205.
26. Dadamouny, M.A., Zaghloul, M.S., & Moustafa, A.A. (2012). "Impact of Improved Soil Properties on Establishment of Moringa peregrina seedlings and trial to decrease its Mortality Rate.". Case Study. Egyptian Journal of Botany, 52(1), 83-98.
27. Krotz, Dan. "New Study: As Climate Changes, Boreal Forests to Shift North and Relinquish More Carbon Than Expected | Berkeley Lab". News Center. Retrieved 2015-11-09.
28. Solomon, S., et al. (2007). Technical Summary. In 'Climate Change 2007: The Physical Science Basis. Contribution of Working Group I to the Fourth Assessment Report of the Intergovernmental Panel on Climate Change'. (Eds.

S. Solomon, et al.) pp. 19-91, Cambridge University Press: Cambridge, United Kingdom and New York, NY, USA.

29. Thuiller W; et al. (2008). "Predicting global change impacts on plant species' distributions: Future challenges". Perspectives in Plant Ecology, Evolution and Systematics. 9: 137–52. doi:10.1016/j.ppees.2007.09.004.

Index